THE HOMEOWNER'S GUIDE TO
COALBURNING STOVES AND FURNACES

James W. Morrison

ARCO PUBLISHING, INC.
NEW YORK

Dedication

For Wendy with her radiant glow and warm heart for the nice things in life.

Published by Arco Publishing, Inc.
219 Park Avenue South, New York, N.Y. 10003

Copyright © 1981 by Arco Publishing, Inc.

Library of Congress Cataloging in Publication Data
Morrison, James Warner, 1940-

The homeowner's guide to coalburning stoves and
furnaces.

1. Stoves, Coal. 2. Furnaces. 3. Coal. I. Title.
TH7443.M67   697'.042   81-7938
ISBN 0-668-05097-7 (Library Edition)   AACR2
ISBN 0-668-05100-0 (Paper Edition)

Printed in the United States of America

Contents

Acknowledgments

Special thanks are due to Roy Upham, Professor of Chemistry, for his help in describing the physical and chemical properties of coal and for his explanation of its combustion process. Research Librarian Patrice Rafail was most helpful in discovering how coal was used in the early 1900's. We appreciate the response of the many stove manufacturers whose products are shown in these pages and the many state agencies that provided data used in this book. Special thanks go to Dave Lamont, Combustion Engineer for the State of Vermont, for his time, interest and guidance.

James W. Morrison
Manchester, New Hampshire

Chapter 1

Introduction

In North America, coal was used by the Indians even before the European invasion that followed Columbus' exploration. The settlers found plenty of wood and had little need for coal until the early 1800's. By 1900 coal had become America's principal fuel. Coal had every appearance of being the energy source of the future. No other fuel on the horizon (petroleum, natural gas) held the advantages of abundance, ready availability, low cost, and multiple uses.

As petroleum refining expanded in response to the increasing demand for gasoline for use in motor vehicles, fuel oil was an inexpensive by-product that gradually replaced coal for industry and for home heating. The rapid expansion of oil and gas pipelines after World War II brought us even more convenient fuels and replaced coal with oil and natural gas for much of our heating needs. Very little coal was then used for home heating.

This decade has brought about the beginnings of drastic change. The predictions that petroleum and natural gas supplies would run out and the resulting rapid price increases by the Oil Producing and Exporting Countries (OPEC) caused a search for alternatives. Industries and power companies started to consider conversion to coal as a cheaper fuel in greater supply. Homeowners started burning wood in fireplaces, in stoves, and even in furnaces to supplement expensive oil heat and electric heat. More and more others are installing coal stoves and coal furnaces as the economy of coal becomes evident again.

The convenience of liquid and gaseous fuels, and their practical necessity in motor vehicles, have renewed the interest in converting coal to synthetic petroleum (coal liquefaction) and to synthetic "natural" gas (coal gasification). Synthetic fuel processes were developed in Hitler's Germany before World War II. Anticipating mechanized warfare, with its need of fuels

1

for planes, tanks, and trucks, and with no oil fields in Germany, the German chemists and engineers devised two methods of producing hydrocarbon mixtures similar to petroleum. In one method, coal reacts with steam at high temperature and pressure. The carbon of the coal takes oxygen from the water and forms a mixture of carbon monoxide and hydrogen. These, in turn, can react to form a mixture of hydrocarbons that can be separated by the usual petroleum refining processes into gasoline, diesel fuel and other products. In another method, coal reacts directly with hydrogen to produce a mixture that can be refined into gasoline, diesel fuel, fuel oil, and fuel gas.

After World War II, these methods were investigated and improved upon in the United States. One especially interesting development was the perfection of a technique to avoid the mining of the coal entirely by drilling two shafts into a coal seam and burning the coal at the bases of the shafts. The heat formed by the burning coal would cause the surrounding coal to react with the underground water, forming petroleum-like hydrocarbon mixtures. As air was forced down through one shaft to keep the fire burning, the hydrocarbon products were forced out through the other shaft to be refined. This saved the cost of mining and transporting the coal, avoided the hazard to the miners in deep shaft mining, and prevented the insult to the environment of strip mining. These processes were abandoned in the early 1950's because they were too expensive to compete with natural deposits of petroleum and gas. Governments and many industries are returning to these processes of liquefaction and gasification of coal as the economics of energy continue to evolve in the 1980's.

Since World War II, an increasing competition from hydrocarbon fuels has developed in most industrialized countries, first in the United States in the 1950's and then in Western Europe in the 1960's. World coal output today[1] has continued to increase—although at a slower rate than before—owing to rapid growth in the USSR and a recovery of the American coal industry since 1961.

---

[1] In 1970, coal and lignite production in the world was 2965×10^6 metric tons per year or six times the production level in 1900.

Chapter 2

The Petrography
of Coal

Modern civilization is the daughter of coal, for coal today is the
greatest source of energy and wealth

G. Gamician (1913)

Origins of Coal[1]

Coal is a solid fossil fuel which represents energy originating in organic
activity of the geologic past and stored for long ages in the rocks of the
earth's crust. It is a combustible sedimentary rock that had its origin in the
accumulation and partial decomposition of vegetation. Like wood, it is composed of carbon, hydrogen, oxygen, nitrogen, and sulfur, together with
small amounts of other chemical elements. The chemistry of coal is extremely complex, and the proportions of the main elements vary considerably according to the particular kind of coal.

The great coal-making time was the Carboniferous Period, 225 to 300
million years ago. In the Pennsylvanian Period (or the Upper Carboniferous
Period, 300,000,000 B.C.) the climate was subtropical, muggy, rainy, and
frostless around what is today Wilkes-Barre, Pa. In this warm humid climate
there was rapid vegetation growth which provided the vast accumulations
of plant debris necessary to make thick coal deposits. We can anticipate

---

[1]The description and systematic classification of rocks is called petrography.

3

comparable depositional environment of coal today. Peat is presently being made in the marshes and bogs of the Dismal Swamps of Virginia and North Carolina (where modern peat deposits are seven feet thick), the Everglade Swamps of Florida, the lowlands of Ireland, and the deltas and rain forests of the East Indies, Africa, and South America.

The peat deposits in time increase in thickness until they are buried beneath layers of sedimentary deposits of sandstones, lime, and shales. Burial of the peat ends the phase of biochemical change and commences the slow physical and chemical processes leading to formation of coal, i.e., coalification.

As peat deposits are gradually buried beneath increasing thicknesses of younger sediments, the peat is subjected to rising pressures and temperatures (one degree increase for every fifty-five feet of depth) which compress it and reduces its volume by the expulsion of water and volatile compounds. As the resulting enrichment of carbon continues, the peat gradually acquires the solidity, color, and chemical constitution of true coal. Peat increases at a rate of one foot per century, and three feet of composed peat are necessary to make one foot of coal. These changes define a progressive increase in rank of the coal. These geologic formations or cyclothems consist of many layers of sedimentary material; coal usually occurs at the fifth stratum.

The Rank Classification of Coal

The term "fuel ratio" refers to the ratio of fixed carbon to volatile matter. Beginning with low-volatile matter and high-moisture coals, there are several classes of coal: peat, lignite, sub-bituminous, low-rank bituminous, medium-rank bituminous, high-rank bituminous, low-rank semi-bituminous, high-rank semi-bituminous, semi-anthracite, anthracite, and super-anthracite.

The classes given above can be placed into four major groups, although the placing of boundaries between them is arbitrary.

Peat, a SUB-BITUMINOUS COAL, is more or less decomposed spongy plant debris with very high water content (90 percent by weight of water) and a relatively low heating value. It will burn when dry, emitting a great amount of smoke. Peat is used to heat the homes of Ireland and Scotland, and improves the taste of their whiskies.

BROWN COALS AND LIGNITES represent an early stage of coal formation. Lignites still contain recognizable woody material (i.e., lignin; hence the name), but brown coals do not. Both types have high water content and contain appreciable quantities of volatile constituents (between 40 and

55 percent). In time and with the continued application of heat and pressure, the process of coalification transforms lignitic coal into higher-rank coals of lower moisture content.

BITUMINOUS COALS are the most widely distributed and best known of the solid fuels. Bituminous coals have been further refined by pressure and heat and contain a higher percent of carbon and lower amounts of water than lignites. The name comes from their tendency to burn with a smoky flame and to melt when heated in the absence of air. Bituminous and sub-bituminous coals with volatile contents of 30 to 45 percent are widely used for steam-raising and general combustion, while those containing 20 to 30 percent volatiles provide the coke required for the iron and steel industries and the smokeless solid-fuel market. It should be noted that coke is produced by driving off the volatile content of soft bituminous coal by heating it in the absence of air. Since most of the smoke from soft coal comes from volatile matter, coke can be burned with little or no smoke. The gaseous component is, of course, coal gas, which supplied virtually all fuel gas requirements before natural gas. Coal gas is mainly methane (CH_4), but contains other hydrocarbons, as well as hydrogen and carbon monoxide.

Ordinary household coals are normally of bituminous type (soft coal), characterized by bands of "bright" and "dull" constituents (look at a piece of coal if you have some). The "bright" bands are derived mainly from wood or bark material, in which original cellular structures are often preserved, while the "dull" bands are formed from degraded plant debris, including plant spores or pollen.

A certain proportion of inorganic mineral matter is always present, derived from the soils upon which the original plants were growing, and from mineral salts in the plant tissues. This is the main constituent of the ash remaining after coal is burnt.

Cannel coal is a soft coal with a high percentage of volatile matter that burns with a hot candle-like flame. It can be used in a fireplace but not in a furnace or stove. It contains substantial amounts of volatile materials which tend to expand when heated. If burned in a confined area, such as a stove, the fire will be too big and too hot to be readily controlled and small explosions of the volatile matter may occur.

ANTHRACITE (hard coal) is the type recommended for home heaters. It burns freely and uniformly with a short, blue flame and produces little if any smoke when properly burned. Anthracite is more desirable than bituminous (soft coal) for use in domestic stoves and furnaces as soft coal creates dust when handled and produces large amounts of smoke and soot when burned at a slow rate.

For homeowners, anthracite is a natural fuel that is inherently smokeless under all heating conditions of use. Anthracite will not produce smoke in any type of heater, irrespective of the method of firing, the condition of equipment, or carelessness on the part of users. Anthracite also has the advantage of being less affected by lack of skill in firing than competitive fuels.

Coal rank increases with depth of burial and, since the changes are irreversible, the rank of a coal seam indicates the maximum temperature and pressure at which it was formed, even though great thicknesses of overlying strata may since have been removed by erosion. Additional pressures associated with folding movements may also increase the rank of coal. In Pennsylvania, for example, after the Carboniferous Period, there were earth upheavals and mountain building which applied more pressure to the coal as the land buckled and seams were folded with faults.

The geology of coal deposits shows that coal seams are relatively thin sheets interbedded with other sedimentary rocks and varying in thickness from mere films to several feet.

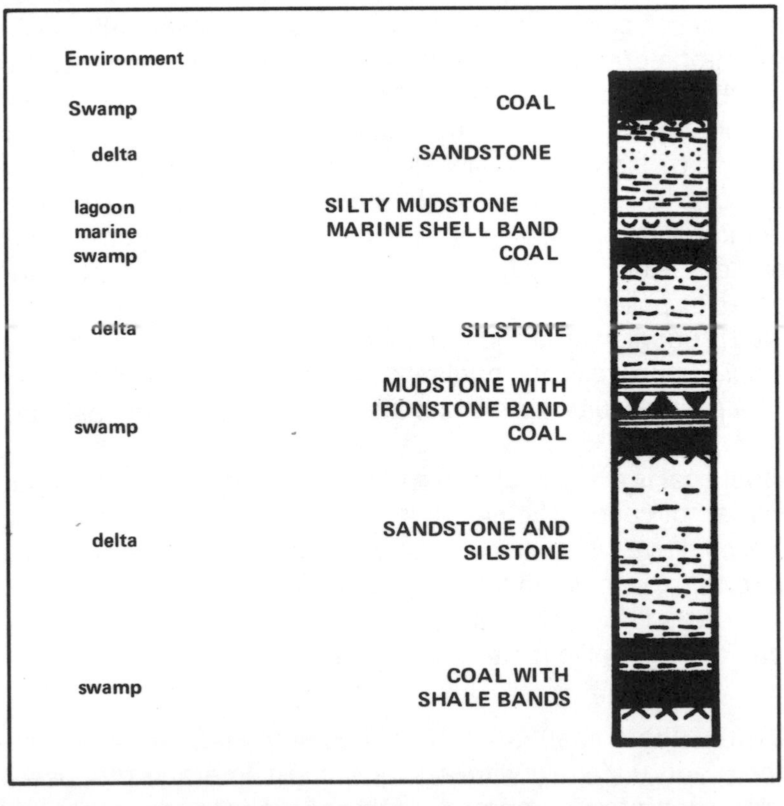

Environment		
Swamp	COAL	
delta	SANDSTONE	
lagoon	SILTY MUDSTONE	
marine	MARINE SHELL BAND	
swamp	COAL	
delta	SILSTONE	
	MUDSTONE WITH	
	IRONSTONE BAND	
swamp	COAL	
delta	SANDSTONE AND	
	SILSTONE	
swamp	COAL WITH	
	SHALE BANDS	

Fig. 1—Idealized diagram illustrating a typical coal seam sequence from the Carboniferous Period.

The ratio of total thickness of a coal seam to that of the enclosing strata is commonly very small, i.e., 1:10 or 1:100. The non-coal strata comprise shales, mudstones, siltstones, and sandstones. Sedimentary structures of crossbedding, ripple marks, and buried channels demonstrate that the sediments were deposited in shallow water as one might see from rapid flooding of the peat swamps.

World Energy and Coal Reserves

Over the last 20 years, the world demand for oil has risen very rapidly while coal production has grown slowly. Oil has become a primary source of energy in almost every country. Simultaneously, the share of oil production of those countries now comprising OPEC has risen to over half of the world total.

Most projections of world energy production indicate a slower (but still considerable) increase in oil use and slow growth in coal production.

The U.S. has larger coal reserves than any other country, and they can be mined with relative ease. Major competitors of the U.S. in the world coal market are the Australians, Canadians, Poles, South Africans, and (soon) the Chinese.

Coal is the major energy reserve of the U.S.—although the size of the reserve is not a good indicator of the relative cost of producing each fuel. Reserve estimates such as these always reflect a judgment about economics—more or less oil, coal, or gas is recoverable depending on the price it brings. Recoverable reserves of coal as used here are 50 percent of demonstrated recoverable reserves (known or indicated deposits which are economically recoverable in the present technology). In practice, more than 50 percent of such deposits are often recovered. Data for petroleum, natural gas, and NGL also represent proved reserves economically recoverable with present technology. Data for shale oil and oil in bituminous rocks include measured, indicated, and inferred reserves and are not wholly comparable with the estimates for other fuels.

While projections vary, coal use is expected to grow to roughly 1 billion tons in 1985 and 1.1–1.3 billion tons by 1990. Seventy-seven percent of current U.S. coal consumption is used in the generation of power in the utility sector; utilities are expected to continue to be the major component of demand. Between 1976 and 1990, utility coal demand is projected to more than double.

The ten-year historical decline in industrial coal use is expected to reverse in the near-term in response to higher oil and natural gas prices. Implementation of mandatory industrial coal conversion programs may further accelerate the expected growth in industrial coal use. Although the tonnage of non-coking (steam) coal used by the industrial sector may double by 1985, that sector's share of coal use is projected to be stable through 1990.

RECOVERABLE* COAL RESERVES — WORLD
(1974 estimate)

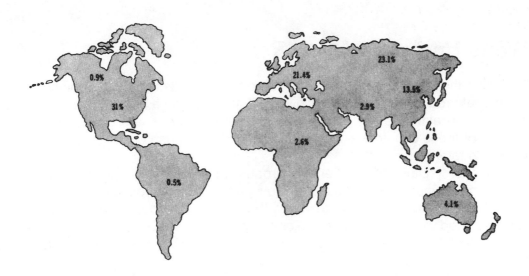

		Approximate Tonnage, Millions	Percent of Total World Reserves
NATION	United States	200,379	31%
	U.S.S.R	150,576	23.1%
	China	88,185	13.5%
	Canada	6,103	0.9%
AREA	Europe	139,746	21.4%
	Oceania	27,027	4.1%
	Rest of Asia	19,354	2.9%
	Africa	17,227	2.6%
	Latin America	3,089	0.5%
WORLD TOTAL		651,686	100%

* Economically recoverable reserves; amount of reserves in place that can be recovered under present local economic conditions using available technology.

Source: World Energy Conference: Survey of Energy Resources 1974.

Fig. 2

U.S. coal is found in 31 states and mined in 26 of them. Coal reserves are adequate to support several hundred years of production at current rates. Eastern coal reserves are primarily deep (i.e., underground), mine-able, and of high quality. The seams are fairly thin—from three to six feet, generally—and may have from 0.7 percent to up to 4 percent sulfur content. Midwestern coal reserves include more surface-mineable coal, almost all high in sulfur content. Seams are often four to ten feet thick. Western coal reserves include vast amounts of surface-mineable, thick (20–100 feet), low-sulfur coal.

Bituminous coal, found in all U.S. regions, has a high heat content—11,500–13,500 Btu per pound—and a variable sulfur content. Some bituminous coals are used to make coke for blast furnace smelting of iron. Sub-bituminous coal, found mainly in the West, has a heat content of 8,000–10,000 Btu per pound and usually a low sulfur content. Lignite (brown coal) has a low heat content (around 7,000 Btu per pound), is found in the West and Gulf Coast regions, and cannot easily be transported long distances or stored due to its combustibility. Anthracite coal (around 13,000

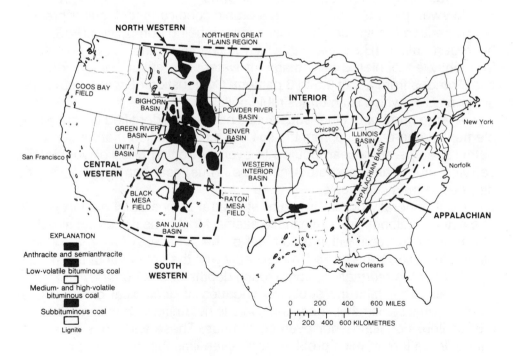

SOURCE: P. Averitt, *Coal Resources of the United States, Jan. 1, 1974*, U.S. Geological Survey Bull, 1412, at 5 (1975).

Fig. 3—Coalfields of the Conterminous United States

Btu per pound) is produced mainly in Pennsylvania, and is a high-heat-content, low-sulfur coal used mainly in small residential and commercial furnaces and boilers.

Types of U.S. Coal

The composition, heat content, and locale of typical ranks of coal are shown in Fig. 3. High-volatile bituminous coals have agglomerating characteristics (tendency to fuse when heated), as do higher ranks of coal generally. The lowest rank coals, sub-bituminous and lignite, are nonagglomerating. In general, moisture and volatile content decrease and fixed carbon increases with increased rank.

Boilers and furnaces can be designed to burn almost any coal, but may not be readily adaptable to other coals even of the same types. The characteristics that affect equipment design are the heating value, ash melting characteristics, hardness of both the coal and its mineral matter (for grinding), and mineral composition (which affects corrosion and fouling tendencies). Coal beneficiation can improve undesirable characteristics and reduce the differences among coal types.

In swamps and bogs, under anaerobic conditions, any sulfur present will be reduced to sulfide, the most common being iron pyrites (FeS_2) and hydrogen sulfide (H_2S), the gas which accounts for the rotten egg smell frequently encountered near stagnant water. The sulfur content of most coals ranges from about 0.5 to 3 percent. Combustion of coal releases the sulfur as SO_2, which is an undesirable pollutant.

A large number of other substances have been identified in coal. Many elements were drawn up from the soil into the vegetation that formed coal. Later, streams and ocean poured in over the rotting beds of vegetation, carrying mud, sand, and other minerals. Thus, coal can contain varying degrees of many elements.

Coal generally occurs in major structural basins as shown in Fig. 3. The actual coalbeds are within these basins. Most beds are broad and thin, with most of the coal within 3,000 feet of the surface. A few in the Rocky Mountain region were canted very steeply with the upheaval of the mountains during their formation. Some reach depths of 30,000 feet.

Coalbeds exhibit the structural vagaries of geological changes that have occurred since their inception. Coal is not uniform in thickness in the roof or floor condition—nor even continuous. These variations present a changing pattern of mine problems and even limit the amount of reserves that can be safely and economically recovered. Other beds, however, are startlingly level and geometrically uniform for miles in every direction. In the Appalachian region, an area of ancient eroded mountains, outcrops can often be found circumscribing adjacent mountains. Some beds are split into several layers with intervening rock layers. Beds quite unrelated geologically can be stacked atop each other.

The most desirable commercial beds are those that are thick, uniform, and very extensive. The Pittsburgh bed is mineable over an area of 6,000 square miles in Maryland, West Virginia, Pennsylvania, and Ohio, with a thickness of up to 22 feet. The edges thin out over a much broader area. This bed presents a striking continuity over this vast region and has yielded more than 20 percent of the coal mined in the U.S. to date. Other Eastern beds also cover several thousand square miles, with coal two to ten feet thick. Midwestern beds can be even thicker and broader. Herrin (No. 6) in Illinois, Indiana, and western Kentucky covers 15,000 square miles, with a thickness of up to 14 feet. The largest bed in the United States is the Wyodak in the Powder River Basin in Wyoming and Montana. Thicknesses of 50 to 100 feet, and 150 feet have been observed.

Coal Distribution

Coal deposits are dispersed over much of the country. The major coal provinces are shown in Fig. 3. About 13 percent of the entire country has coal beneath the surface. About two-thirds of West Virginia and Illinois are above coalbeds, while more than 40 percent of North Dakota and Wyoming are above coalbeds. The deposits vary widely in size and quality.

Appalachian Region

Coal production in the Appalachian region may be characterized as being in the advanced stages of maturity (i.e., having supplied 60 percent or more of total national production annually since coal mining began in the United States). The coal is obtained from a large number of mines of widely varying sizes and from a large number of coalbeds, many of which occur in areas of steep terrain.

The remaining recoverable underground reserves are large, but substantial portions may remain unmined because they occur in uneconomical seams. Surface mine reserves are probably insufficient to support even present production rates for more than a decade. The coal is generally the highest grades of bituminous, but it often has a high sulfur content.

Eastern Interior Region

Coal production in the eastern interior region involves a small number of individual mines with relatively high annual production rates. Indiana, western Kentucky, and Illinois have annually supplied from 20 to 25 percent of total national production for many years.

The greatest portion of the total remaining recoverable reserves for underground mining occurs in beds underlying already mined-out seams. Very large quantities remain, especially in Illinois.

Surface mine production has remained essentially constant for a number of years. As surface topography is generally favorable for large-scale surface mining, it is likely that this method will be pursued further (although at reduced annual rates from individual mines) in thinner beds as reserves in the principal beds become exhausted.

Prospects are favorable for substantial expansion of production from underground mines. Surface mines should be able to maintain their present rate for many years. The coal tends to have a slightly lower heat content than Appalachian coal, and the sulfur content can be quite high.

Western Interior Region

Underground production has entirely ceased in Arkansas, Kansas, and Oklahoma and has been very limited for many years in Iowa. While surface production has remained approximately constant in recent years, the amount produced is not significant. The coalbeds are persistent but thin throughout the region, except in Iowa and the southern portion of Oklahoma, where some beds are of moderate thickness. Were it not for local markets, it seems doubtful that much coal would be mined in this region. Some coal occurring under regions with difficult mining conditions in the southern portions of Oklahoma and Arkansas is of metallurgical and foundry grade and is being mined and shipped out of the region. Production in this region is unlikely to increase significantly in the near future.

Surface-Mineable Regions

The western states contain the most recoverable surface-mineable reserves (75 billion tons) and are virtually untouched (except for a few large active operations). The beds generally are thicker than the beds being surface-mined elsewhere in the country (excluding Pennsylvania anthracite). In Washington and Arizona, surface-mineable reserves are essentially those remaining in the surface mines now active in each state; in South Dakota and Utah, surface-mineable reserves are relatively small, and in Colorado, surface-mineable reserves are largely confined to the two northwestern counties.

The sparsity of surface mining in these states has been due to their remoteness from large markets. In anticipation of future need, intensive prospecting and acquisition by lease or purchase was initiated about ten years ago, and many large individual mines already are operating or have been planned. Contracts for large annual shipments from some of these unit areas to existing or contemplated new powerplants have been negotiated, while other units have been set aside for prospective gasification or liquefaction plants. Plans for new or connecting transportation facilities have been completed or are being developed.

Underground Reserves

Excluding Washington, New Mexico, and Alaska (where significant recovery of underground reserves is considered dubious), and North Dakota and Texas (where the lignite beds are considered unsuitable for underground mining), the western states of Colorado, Utah, Montana, and Wyoming contain more than 69 billion tons of recoverable underground reserves. Although important portions of the coalfields of Utah, Colorado, and Wyoming have undergone substantial depletion, many relatively unmined areas remain. Some of these may be comparatively difficult to mine because of the degree of dip or the excessive thickness of overlying cover, but in many areas mining conditions are favorable.

U.S. ENERGY SUPPLY

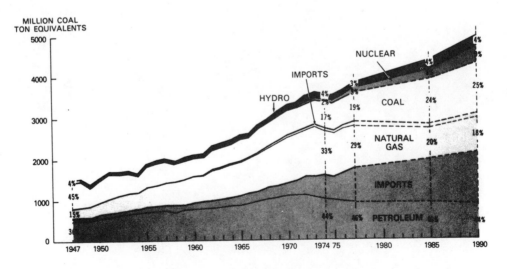

Note: Percent figures represent percent shares of total supply.

Fig. 4—U.S. Energy Supply

Significant increases in underground production in Utah and Colorado can be expected in the next few years. Similar interest in the underground reserves in Wyoming and Montana may also develop, although underground recovery of as much as 50 percent of the thicker beds may be difficult with present technology, which is generally limited to bed thicknesses not exceeding 15 feet. If or when additional production becomes necessary, there is no reason why underground production could not be conducted simultaneously or subsequent to surface mining.

Most western coal is sub-bituminous or lignite. It has a substantially lower heat content and a higher moisture content than eastern coal.

U. S. reliance on coal has declined from nearly one-half of our energy supply at the end of World War II to less than 20 percent today. Oil and natural gas have grown in relative importance during the same period; the share of energy provided by cheap gas supplies has doubled. Forecasts generally show a gradual increase in the production of coal. This should be good news for anthracite coal users.

Chapter 3

The Characteristics of Coal

Dirty British coaster with a salt-caked
   smoke stack,
Butting through the Channel in the mad
   March days.
With a cargo of Tyne coal
Road-rail, pig-lead,
Firewood, iron-ware, and cheap tin trays.
             Masefield, *Cargoes,*
                Stanza 3

Ultimate Analysis

The method of ultimate analysis requires a chemical determination by weight of each of the major chemical elements of which coal consists: fixed and combined carbon, hydrogen, oxygen, nitrogen, sulfur, water, and ash. The substances that burn and have heating value are carbon, hydrogen, and sulfur.

The ultimate analysis is needed to determine the combustion air requirements, the properties of the products of combustion, and the size of air and flue gas-handling equipment, and to make complete heat-balance tests of the steam-generating unit. It is also important in the selection of coal for various special-purpose applications.

The ultimate analysis by the Bureau of Mines of the same 2-inch by 1¼-inch bituminous nut coal for which the proximate analysis was given follows:

TABLE 1

	As Received	Dry	Moisture and Ash Free
Hydrogen	5.5	5.0	5.5
Carbon	66.5	72.9	80.7
Nitrogen	1.4	1.5	1.6
Oxygen	15.1	8.0	9.0
Sulfur	2.7	2.9	3.2
Ash	8.8	9.7	
	100.0	100.0	100.0
Btu	11,940	13,100	14,500

In the "as received" analysis the moisture has been broken down into hydrogen and oxygen and combined with the free hydrogen and oxygen. This explains the decrease in both hydrogen and oxygen when the analysis is converted to a "dry basis." The Bureau of Mines procedure in reporting ultimate analysis is in accordance with ASTM standards; however, some commercial laboratories report moisture separately.

Heating Value of Coal

The heat value is the total amount of heat liberated by the complete combustion of a unit weight or volume of solid fuel. The heat value determination is of primary importance to all types of coal users, as it gives a measure of the potential energy contained in the fuel. The heat value is usually reported in terms of British thermal units per pound of coal (Btu/lb.). The Btu determination is made in the laboratory in an apparatus called a calorimeter, where a weighed quantity of pulverized coal is burned in an atmosphere of oxygen under pressure, which in turn heats a measured quantity of water. The quantity of heat liberated is calculated from the rise in temperature of the water, as shown by a very delicate thermometer. The results are reported on a Btu "as received," Btu "dry" or Btu "ash and moisture free" basis.

A practice followed by many users is the purchase of coal on the basis of cost per million Btu. While this provides a common basis or yardstick for the comparison of different coals, it does not take into account other important characteristics of coal that can affect operating efficiency and cost.

The cost per million Btu can be calculated by the following formula:

$$\dfrac{\text{Delivered price per ton in cents}}{\dfrac{\text{Btu per lb. as rec'd.} \times 2000}{1{,}000{,}000}} = \text{Cost per million Btu}$$

If we assume a delivered price of $70.00 per ton for the coal of the analysis given above, we can substitute in the formula as follows:

$$\dfrac{7000}{\dfrac{11{,}940 \times 2000}{1{,}000{,}000}} = \$2.93 \text{ per million Btu}$$

Sulfur

Sulfur is present in the form of organic and inorganic compounds, in varying degrees, as a component of all coals. There are many misconceptions concerning sulfur in coal. It is very important to have reliable information about sulfur. The amount of sulfur is determined by completely oxidizing the coal and converting by a chemical process all of the sulfur contained therein to sodium sulfate. This is chemically separated and the percentage of sulfur calculated and reported as total sulfur.

Total sulfur is composed of three forms: organic, pyritic, and sulfate.

Organic: Sulfur which is present as organic compounds was originally part of the plant material from which the coal was formed. It is present in the organic compounds of the coal matter itself and cannot be removed by any cleaning processes.

Pyritic: Inorganic sulfur occurs in coal as pyrite or marcasite, both of which are ferrous sulfide. Pyrites were formed by mineralization when the coal was being deposited or later was precipitated from highly mineralized water seeping through the coal measures. Pyrites are much heavier than coal and are easily separated from it by washing and cleaning processes.

Sulfate: Sulfate sulfur occurs in coal in almost negligible quantities. It is formed as calcium or magnesium sulfate and may be present as iron sulfate which results from slow oxidation of the pyrite. Sulfate is non-combustible.

Many special processes, such as cement and brick kilns, can and do use coals containing a large percentage of sulfur. Some brick manufacturers depend on the sulfur content to produce color variations in the product. The influence of sulfur and sulfur oxides on the use of coal for steam generation has been much exaggerated. There is no consistent correlation between the sulfur content of coal and corrosion or fouling of heating equipment.

That sulfur in coal increases the clinkering tendency of a given coal is a misconception. Neither organic nor inorganic sulfur has any influence on the softening temperature of the compounds that make up the ash of coal. The iron associated with sulfur, present as pyrite or marcasite, may contribute to the lowering of the fusion temperatures of a coal ash, but the amount of influence the iron has depends upon the other mineral constituents.

TABLE 2
CLASSIFICATION OF COALS BY RANK

Class and Group	Fixed Carbon Limits, percent (Dry, Mineral-Matter-Free Basis)		Volatile Matter Limits, percent (Dry, Mineral-Matter-Free Basis)		Calorific Value Limits Btu per pound (Moist,[a] Mineral-Matter-Free Basis)		Agglomerating Character
	Equal or greater than	Less than	Greater than	Equal or less than	Equal or greater than	Less than	
I. ANTHRACITE							
Meta-anthracite	98			2			
Anthracite	92	98	2	8			nonagglomerating
Semi-anthracite[b]	86	92	8	14			
II. BITUMINOUS							
Low-volatile bituminous coal	78	86	14	22			
Medium-volatile bituminous coal	69	78	22	31			
High-volatile A bituminous coal		69	31		14,000[c]		commonly agglomerating[d]
High-volatile B bituminous coal					13,000[c]	14,000	
High-volatile C bituminous coal					11,500	13,000	agglomerating
					10,500	11,500	
III. SUB-BITUMINOUS							
Sub-Bituminous A coal					10,500	11,500	
Sub-bituminous B coal					9,500	10,500	
Sub-bituminous C coal					8,300	9,500	
							nonagglomerating
IV. LIGNITIC							
Lignite A					6,300	8,300	
Lignite B						6,300	

a. Moist refers to coal containing its natural inherent moisture but not including visible water on the surface of the coal.
b. If agglomerating, classify in low-volatile group of the bituminous class.
c. Coals having 69 percent or more fixed carbon on the dry, mineral-matter-free basis shall be classified according to fixed carbon, regardless of calorific value.
d. It is recognized that there may be nonagglomerating varieties in these groups of the bituminous class, and there are notable exceptions in high-volatile C bituminous group.

Physical Characteristics of Coal

Coals are classified by rank in order to aid in predicting their performance and insure proper selections for their intended use. We previously discussed the major rank groups of lignitic, sub-bituminous, bituminous, and anthracitic. Table 2, Classification of Coals by Rank (ASTM D–388–77), shows the complete classification system. This classification does not include a few coals, principally nonbanded varieties, which have unusual physical and chemical properties and which come within the limits of fixed carbon or calorific value of the high-volatile bituminous and sub-bituminous ranks. All of these coals either contain less than 48 percent dry, mineral-matter-free fixed carbon or have more than 15.500 moist, mineral-matter-free British thermal units per pound.

Coal classified "mineral-matter-free," is not the same as ash-free. During ignition of a coal, changes in the mineral matter occur, causing the weight of the ash to differ from that of the original mineral matter.

The "agglomerating" character of the coal refers to coal which, when heated, passes through a plastic state during which individual pieces fuse together to form a rather solid mass. This is termed "agglomerating," i.e., "caking" or "agglutinating." These terms originated, in part, from tests developed during the past years to determine the suitability of coals for making metallurgical coke. Anthractic coal that is not agglomerating is termed "nonagglomerating" or "free-burning." Agglomerating coals may cause problems in certain home combustion systems.

A number of commercial and industrialized uses of coal are concerned with the following physical characteristics of coal: grindability, fracture, friability, free swelling index, size stability, agglomerating index, carbonizing properties, specific gravity, porosity, color, and luster. These terms are included in the Glossary.

Chapter 4

Combustion of Coal

He works and blows the coals
And has plenty of other irons in the fire
Aristophanes

Coal Combustion Chemistry

In most hand-fired stoves and furnaces, as the coal burns, the incoming air and the gases formed by the combustion are heated (see Fig. 5). These hot gases are lighter than cold air and they rise, passing through the smokepipe and the chimney into the air above. This provides the draft to pull fresh air in through the dampers in the fire door and the ash pit door. The gases rising out of the chimney must be hot so that they will rise and provide the draft. The amount of draft can be controlled with a check damper as also shown in Fig. 5. The wider the check damper is opened, the more air enters through it and the less air enters through the fire.

Complete combustion occurs when all of the fuel is oxidized by the air. Perfect combustion occurs where there is no excess air in the combustion chamber after complete oxidation has occurred. Except under controlled laboratory conditions, however, a continuous perfect combustion process is unattainable with solid fuels. Therefore, some excess air is necessary to assure complete combustion. For that reason, stoves and fireplaces are designed to operate with considerable excess air to prevent intermittent periods of incomplete combustion. (See Fig. 6.)

The volatile gaseous matter which is initially driven off coal by heat represents about one-third of the total heat value in coal, while the coke,

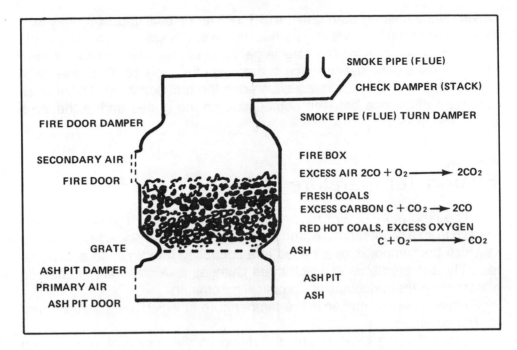

Fig. 5

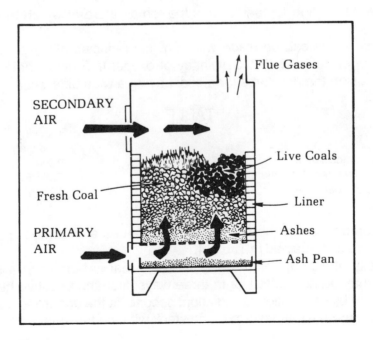

Fig. 6

which is incompletely burned in the fuel bed, represents approximately two-thirds of the total heat value. The heat value of this coke is released in the form of CO, which joins the volatile gases to produce some two-thirds of the heat in the combustion space or area over the fuel bed. The other one-third of the heat value remains situated in the fuel bed itself. There is an important difference between combustion on the grates and in the area above the grates.

Fusion Temperature of Ash

The laboratory determination of ash fusion temperature is made with the ash from a representative sample of the coal. This is placed in a furnace in which the temperature is raised at a controlled rate until the end of the test. The temperatures at which three changes take place are measured either with a thermocouple or an optical pyrometer.

The initial deformation is the temperature at which ash might adhere to cold surfaces.

The softening temperature is defined as the temperature at which clinkering or slagging can be expected. It is this temperature which is usually reported as the ash fusion temperature or ash softening point.

The fluid stage temperature is the temperature at which the ash will flow.

These tests can be made with a mildly reducing atmosphere during the heating (that is, with a deficiency of oxygen). Table 3, below, shows the ash fusion temperatures for a coal having a wide differential:

TABLE 3

	Reducing Atmosphere
Initial deformation	1960°F.
Fusion (softening temperature)	2065°F.
Fluid temperature	2250°F.

The comparison of the various heating values and composition of different coals is contained in Table 4.

Finally, the heat contained in the coal that is burned in a stove or furnace may be accounted for in three ways: first, the effective heat given to the room by convection or radiation; second, in the unburned coal found in the ash; third, what passes up the stack with the flue gas.

TABLE 4
COMPARISON OF THE HEATING VALUES AND COMPOSITION
OF VARIOUS COALS
(Ash-Free Basis)

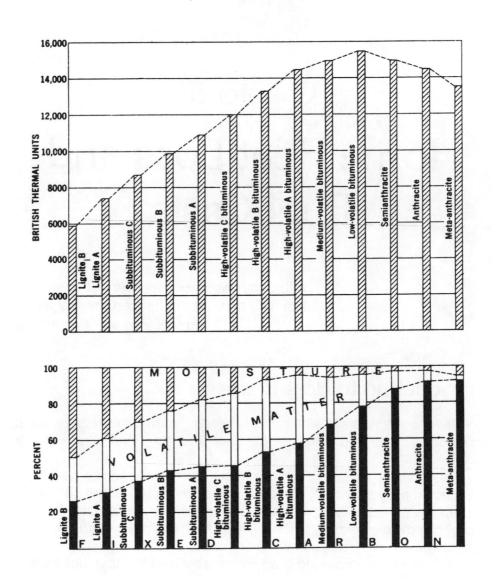

Chapter 5

Home Heat Loss and Fuel Costs

To fireside happiness, to hours of ease
Blest with that charm, the certainty to please.
Samuel Rogers, *Human Life*

Introduction to Heat Loss

Being comfortable in winter means keeping warm. This requires heat, which costs more and more using traditional fuels if one is unwilling to sacrifice comfort.

This chapter explains the facts about comfortable winter heating. It provides an easy method of approximating how much heat will be needed to keep any particular building warm and explains how to assess the benefits of improvements made to the building, such as adding storm windows, insulating exposed areas, and excluding drafts.

We put heat in a house to keep comfortable (and healthy), but the heat passes out of a house to the cold outside surroundings. If we want to keep the building at a comfortable temperature, we must control the actual heat-loss. Remember that the flow of air is always from hot to cold, and cold is really the absence of heat.

The most important fact is that heat always tends to flow from a high-temperature area to a low-temperature area. For example, if you put a pan of cold water on a hot stove, the flow of heat from the stove through the bottom of the pan heats the water up to a higher temperature. Pan bottoms are, therefore, made of materials which conduct heat easily. To keep the pan from losing heat after it comes off the stove, you can stand it on an asbestos pad, a material that resists passage of heat or, in other words, provides insulation. The same principle works with buildings. Take the building below:

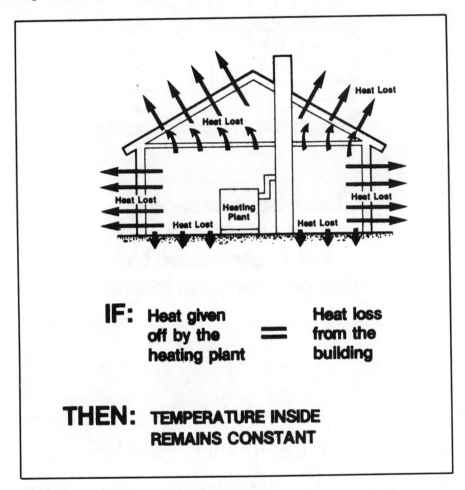

IF: Heat given off by the heating plant = Heat loss from the building

THEN: TEMPERATURE INSIDE REMAINS CONSTANT

Fig. 7

If a building has insulated walls, floors, and ceilings, double-glazed windows, and sealed cracks, then less heat will be needed to maintain comfortable temperatures inside the building. Less heat required means less fuel used, which translates into money saved.

Heat escapes from a building in two ways: by conduction and by infiltration. In the next few pages, these processes will be explained—but first some definitions to make everything clear.

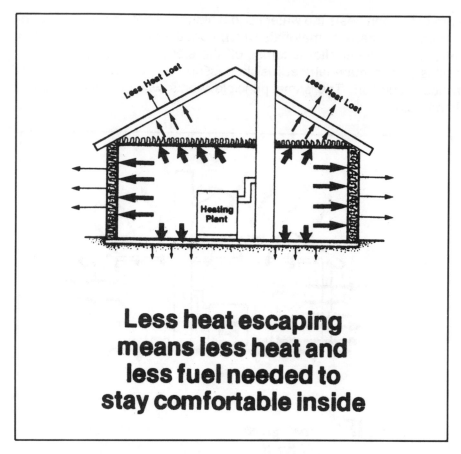

**Less heat escaping
means less heat and
less fuel needed to
stay comfortable inside**

Fig. 8

Useful Terms in Determining Heat Loss

We must have a unit to measure heat losses. Normally we use the British thermal unit—Btu. This is the amount of heat it takes to raise the temperature of one pound of water by one degree Fahrenheit. Another way of defining a Btu is to say it is about the amount of heat given off when a wooden match is burned completely. All fuel values or heat requirements can be expressed in Btu's: for example, using one kilowatt-hour of electricity releases 3,412 Btu's; one pound of wood burned completely will give off about 8,000 Btu's.

Working with Btu's means doing calculations with large numbers in which it is easy to make errors. So, we introduce the term *heating units* to simplify the figuring:

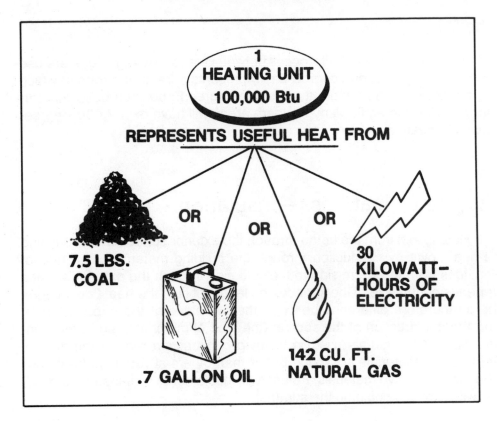

Fig. 9

Obviously, the heating unit is an approximation because not all oil or gas heating plants operate at the same efficiency. However, the heating unit is a fairly accurate estimate of what a normal oil or gas furnace should get out of the quantities of those fuels illustrated above. The heating unit does, in fact, represent about 100,000 Btus.

In this chapter, we calculate heating requirements on a seasonal basis. Therefore, if a building is calculated to require 1,200 heating units, that figure represents the approximate number of gallons of oil it should use per year, if it has an efficient furnace. If the actual fuel use is known and is very different from the calculated figure, a further check needs to be made to find the reason for the discrepancy. It may be due to calculation errors, wrong measurements in the building, or a poorly functioning furnace. It may be that the building has only been partially heated previously, with much of the living space not used in the winter. Working with heating units in this way can enable us to spot errors or circumstances we might otherwise miss. Incidentally, if the furnace that provides heat for the building also provides domestic hot water, this will increase the fuel use by approximately 20 percent.

The Degree-Day

To allow for climatic differences between areas, heating engineers use degree-day figures: one degree-day represents a 24-hour period in which the average outside temperature is one degree F. below a base temperature of 65 degrees F. Many northern areas will have over 7,000 degree-days in a heating season.

Stopping Heat Loss—Insulation

Heat is lost from the home through the exterior surface of the building as heat flows by conduction through the building materials. The rate of heat loss from the warm side to the cold side through the exterior surface depends on the size of the surface, the length of time the heat flow occurs, the temperature difference between the two sides of the exposed area, and the construction of the section (the type of material used in the construction). All materials used in building construction reduce the flow of heat. Some materials are much better than others at reducing heat flow. The more effective materials are used as insulation. A well-insulated building will also stay cooler in the summer.

Types of Insulation

Three general types of insulation materials are commonly used in building construction. They are loose fill, blanket or batt, and rigid insulation.

Loose fill include such types of insulation as glass, rockwool, cellulose fibers, and wood fibers. Fill type insulating materials are best utilized on horizontal surfaces, such as ceiling areas. This type of insulation used in vertical areas tends to settle, and unless provision is made to refill the space, cold spots can occur.

Blanket, or *batt insulation,* is commonly made of glass, rockwool, or wood fiber. It is usually enclosed in a paper envelope or fastened to a backing of kraft paper or aluminum foil. Some blanket types of insulation have no backing and are intended to be used when no vapor barrier is required. Blanket insulation comes in rolls of various lengths and thicknesses. Batt insulation is usually thicker and comes in shorter lengths. Both blanket and batt insulation are available for framing spacing of 16 and 24 inches. Other widths are available on special order.

Rigid insulation, in addition to providing insulating value, also provides structural strength. Rigid insulation is available in board form, such as various fiberboard materials and foamed plastics. Rigid insulation is used quite extensively by contractors and not individual homeowners. In some instances, this type insulation is less expensive.

Table 5 lists the insulating value of most of the common material found in house construction. The R value shown in the right-hand column indicates the effectiveness, or resistance value of the material. *The higher the resistance value, the better the insulating quality.* When building sections are made of several materials, the resistance value of each of the individual materials can be added together to obtain the overall total resistance value. Once you know the overall R value you can use it to determine heat loss. Thicknesses of 3½ inches (R-11) and 6 inches (R-19) are most common.

Vapor Barriers

In the winter, moisture moves from the inside of the home to the outside through the exterior surfaces. Vapor barriers are installed to reduce the flow of moisture through the insulation so that condensation will not occur. Blanket or batt insulation usually has vapor barriers attached. Polyethylene film (four mils thick) can be used as a separate vapor barrier if needed. Vapor barriers should always be installed on the warm side (inside) to stop the moisture before it reaches the insulation. If possible, vent the cold side of the insulation to the outside to remove moisture which escapes through the insulation. When a blanket or batt insulation having an attached vapor barrier is used, kraft paper backing is usually cheaper than foil backing. If foil backing is used, a ¾-inch to 3-inch clearance is necessary in order to maintain effectiveness.

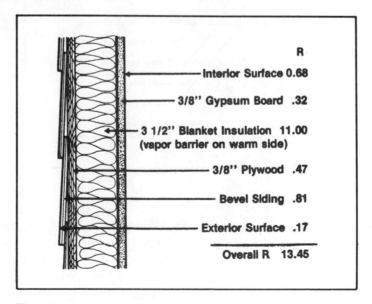

Fig. 10

Exterior Walls

To determine the insulating value of an exterior wall section, it is necessary to know the construction of the wall. Using Table 5, determine the R value for each material making up the wall. Add together these values to obtain the overall R value of the wall.

Ceilings and Roofs

The insulating value of roof and ceiling sections can be determined by adding the R value of each of the materials making up the section. It is necessary to know the construction of the ceiling or roof section. Add together the R values of the materials making up the section from the values given in Table 5 to obtain the overall R value of the section. For ceilings having attic space over the insulation, use an interior surface resistance for the surface next to the attic due to the fact that still air conditions exist on the outside of the insulation. The illustration shows the procedure for determining the overall resistance value for ceiling sections.

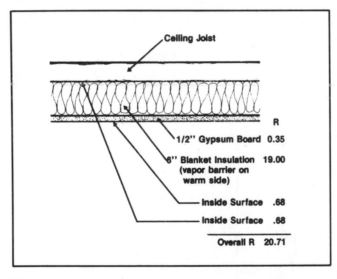

Fig. 11

For roof sections, the procedure to determine overall resistance value is similar to that for the wall section. First, determine the construction of the roof section, and then add the resistance of the individual materials making up the roof section to obtain the overall R value for the roof. The accompanying diagram shows the procedure for determining the overall R value for roof areas.

TABLE 5
Insulation Value of Common Materials

MATERIAL	THICKNESS (Inches)	R VALUE
Air Film and Spaces:		
Air space, bounded by ordinary materials	¾ or more	.91
Air space, bounded by aluminum foil	¾ or more	2.17
Exterior surface resistance	—	.17
Interior surface resistance	—	.68
Masonry:		
Sand and gravel concrete block	8	1.11
	12	1.28
Lightweight concrete block	8	2.00
	12	2.13
Face brick	4	.44
Concrete cast in place	8	.64
Building Materials—General:		
Wood sheathing or subfloor	¾	1.00
Fiber board insulating sheathing	¾	2.10
Plywood	⅝	.79
	½	.63
	⅜	.47
Bevel-lapped siding	½ × 8	.81
	¾ × 10	1.05
Vertical tongue and groove board	¾	1.00
Drop siding	¾	.94
Asbestos board	¼	.13
⅜″ gypsum lath and ⅜″ plaster	¾	.42
Gypsum board (sheet rock)	⅜	.32
Interior plywood panel	¼	.31
Building paper	—	.06
Vapor barrier	—	.00
Wood shingles	—	.87
Asphalt shingles	—	.44
Linoleum	—	.08
Carpet with fiber pad	—	2.08
Hardwood floor	—	.71
Insulation Materials (mineral wool, glass wool, wood wool):		
Blanket or batts	1	3.70
	3½	11.00
	6	19.00
Loose fill	1	3.33
Rigid insulation board (sheathing)	¾	2.10
Windows and Doors:		
Single window	—	approx. 1.00
Double window	—	approx. 2.00
Exterior door	—	approx. 2.00

Source: **ASHRAE** Guide and Data Book.

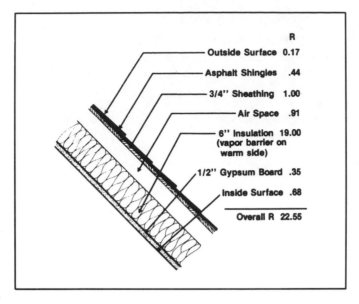

Fig. 12

Floors

To determine the insulating value of floors, add the R value of the individual materials making up the floor section together to determine the overall R value. Use the interior surface resistance for the surface next to

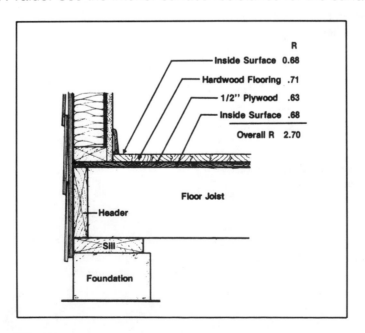

Fig. 13

the basement or crawl space area. The heat loss from floors depends on the temperature below the floor. Basement and crawl space temperatures depend on the quality of construction. In calculating floor heat loss in this manual, a floor exposure factor is used to estimate changes in floor heat loss due to different types of foundation construction.

Building Heat Loss by Infiltration

Any building will constantly exchange air with its environment: outside air leaks in, inside air leaks out. A certain amount of this exchange (say, one complete air change per hour) is necessary for ventilation, but most buildings have much more than is needed. In winter, the air that leaks in is cold; the air that leaks out is warm; fuel is used to supply this temperature difference. Exfiltration (the flow of air is always from hot to cold) is the main reason for a cold house.

This leakage or infiltration is caused by wind, the building acting as a chimney, and the opening of outside doors.

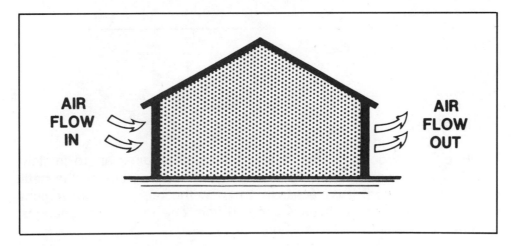

Fig. 14—Infiltration by wind

The effect of door openings and wind needs little explanation, but the chimney effect may not be obvious. When air in a building is warmer than the outside air, the entire building acts like a chimney—hot air tends to rise and leak out of cracks at the upper levels and sucks cold air in through cracks at the lower levels. Both the temperature difference and building height contribute to this effect. A two-story house having a 68°F. inside temperature and a 30°F. outside temperature will produce a "chimney" leakage equivalent to a ten-mile-per-hour wind blowing against the building.

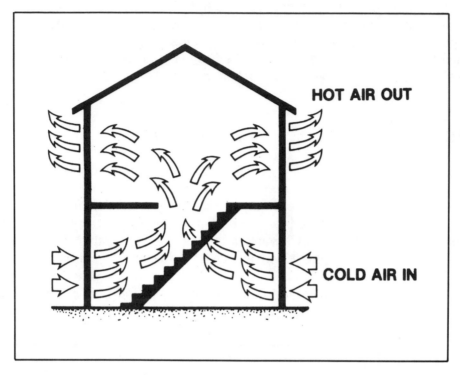

Fig. 15—"Chimney Effect"

Exfiltration

Each cubic foot of air that enters the house requires approximately 0.02 Btu to raise the temperature one degree F. To determine the heat loss from infiltration, it is necessary to know the rate of air movement through the home. Most houses undergo from one to three air changes per hour, depending on construction.

Comparing Fuel Costs

The four fuels commonly used for domestic water-heating are wood, coal, oil, and gas. Electricity, though not a fuel, is being increasingly used. Each of the four fuels should be considered separately.

The therms of heat per dollar should not be the sole consideration in selecting the heating fuel. Installation cost, the efficiency with which each unit converts fuel into useful heat, and the insulation level of the house should also be considered. For example, electrically heated houses usually have twice the insulation thickness, particularly in the ceiling and floor and,

HOME HEAT LOSS AND FUEL COSTS

therefore, may require considerably less heat input than houses heated with fuel-burning systems. To compare costs for various fuels, efficiency of combustion and heat value of the fuel must be known.

Heating units vary in efficiency, depending upon the type, method of operation, condition, and location. The efficiencies in utilizing fuels given in this book are recognized by the American Society of Heating, Refrigeration and Air Conditioning Engineers (ASHRAE) as being reasonable values where the heating equipment is properly installed and adjusted and in good condition. Stoker-fired (coal) steam and hot-water boilers of current design, operated under favorable conditions, have 60 to 75 percent efficiency. Gas- and oil-fired boilers have 70 to 80 percent efficiency. Forced-warm-air furnaces, gas-fired or oil-fired with atomizing burner, generally provide about 80 percent efficiency. Oil-fired furnaces with pot-type burner usually develop a maximum 70 percent efficiency.

Fuel costs vary widely in different sections of the country, but oil, gas, and wood are less desirable with recent price increases. However, for estimating purposes, the data given in Table 6 can be used to figure the comparative costs of various fuels and electricity based on local prices. Here the efficiency of electricity, gas, oil, and coal is taken as 100, 75, 75, and 65 percent, respectively. The efficiencies may be higher (except for electricity) or lower, depending upon conditions; but the values used are considered reasonable. The heat values are taken as 3,413 Btu per kilowatt-hour of electricity for resistance heating; 1,050 Btu per cubic foot of natural gas; 92,000 Btu per gallon of propane (LP) gas; 139,000 Btu per gallon of No. 2 fuel oil; and 13,000 Btu per pound of coal. Remember, a therm is 100,000 Btu.

TABLE 6
FIGURING COMPARATIVE COSTS OF VARIOUS FUELS AND ELECTRIC ENERGY TO SUPPLY ONE THERM OF USABLE HEAT

Fuel or energy	Quantity to supply one therm usable heat	Multiply values in column (2) by local unit costs	Comparative costs per therm of heat supplied to living space in cents
(1)	(2)	(3)	(4)
Coal	11.8 lbs. = .006 ton	per ton	————
Electricity	29.3 Kw-hr	per Kw-hr	————
Fuel oil			
No. 2	.96 gal	per gal	————
Gas, natural	127 ft.	per ft	————
Gas, LP			
(propane)	1.45 gal	per gal	————

The use of wood requires more labor and more storage space than do other fuels. However, wood fires are easy to start, burn with little smoke, and leave little ash.

Most well-seasoned hardwoods have about half as much heat value per pound as does good coal. A cord of hickory, oak, beech, sugar maple, or rock elm weighs about two tons and has about the same heat value as one ton of good coal.

Heat values of the different sizes of coal vary little, but certain sizes are better suited for burning in fire-pots of given sizes and depths. Both anthracite and bituminous coal are used in stoker firing. Stokers may be installed at the front, side, or rear of a furnace or boiler. Space must be left for servicing the stoker and for cleaning the furnace. Furnaces and boilers with horizontal heating surfaces require frequent cleaning, because fly ash (fine powdery ash) collects on these surfaces. Follow the manufacturer's instructions for automatic operating stokers.

Chapter 6

Selecting a Proper Stove

In fleeing the ashes he's fallen
into the coals
                    Apostolius: Proverbs

Types of Stoves and How They Heat

The design of a coal stove in any given category may vary significantly from stove to stove, yet the basic principle remains that of extracting heat from the fire. There are thousands of manufacturers of coal-burning stoves throughout the world. Appearance, style, finish, construction, materials, and weight are some of the characteristics which need to be evaluated. Durability of welding, sharpness of cabinet or stove edges which may scratch or cut people, and ability to burn coal efficiently for maximum heat are other factors to consider.

Low-efficiency stoves (20–30 percent efficient) are technically the simplest. They have a straight air path either across or through the fire, allowing sufficient air for combustion of the coal (primary air) with no accommodations for additional air to support the flame combustion (secondary air). Simple box stoves, Franklin stoves, pot belly stoves, parlor stoves and sheet metal stoves are representative of this category. These stoves are not airtight, and may burn bituminous or even cannel coal.

Medium-efficiency stoves (35–50 percent efficient) are more sophisticated. They have better control of the amount of primary and secondary air and are built without excessive air leaks. These stoves are commonly called airtights. Air flow to the fire is controlled to insure maximum utilization of the burn. A thermostat is commonly utilized to ensure a constant burning rate.

High-efficiency stoves (50–60 percent efficient) are the most technically sophisticated. They regulate air flow as in the simpler airtights but employ baffles, long smoke paths, and heat exchangers to extract as much heat as possible from the fire. These stoves are the most expensive to buy but are the cheapest to operate as they deliver the most heat per unit of fuel.

Most coal stoves transfer heat to the room by radiating heat from the hot surface of the stove. Radiant heaters produce heat that is most intense in close proximity and diminishes rapidly with distance from the stove. Surfaces in direct line with the stove will be heated.

The following are some representative types of stoves.

Fig. 16—Chubby Moe from the All Nighter Stove Works, Inc.

Circulating stoves are constructed with a metal box spaced about one inch from the wall of the firebox. Vents in the top and bottom of the outer box allow natural air currents to carry the heat away from the stove. The outer surface of a circulating stove is not as hot as that of a radiant stove and thus it can be installed closer to combustible material than a radiant stove. Circulating stoves are better suited to heating a large room than radiant stoves.

Fig. 17—Franco-Belge Coal Circulating Stove

The combination type stove is similar to the Franklin stove in its use. It can be operated as an open fireplace or a closed stove. Most of these are manufactured of cast iron and are large enough to heat one or two rooms. Some of the stoves are built with airtight doors and good draft control.

A *box stove* is just what the name indicates—a box. It can have either a square or rectangular cross-section and is most often supported on legs.

Fig. 18—Home Heater Model #1

The coal is placed on grates. These stoves are available in many sizes, of airtight construction and with baffles or warming boxes to increase their efficiency.

The *parlor stove* was designed to heat a single room. It usually has a double steel wall or jacket around it which gives a lower surface temperature than a single-wall stove. It is often available with a thermostat that regulates the drafts and thereby the temperature in the room.

The *pot belly stove* is most often associated with railroad stations and country houses. It is usually made of cast iron and often has ornate designs and trim. The small-diameter firebox requires small amounts of nut coal.

Fig. 18A—Comforter Coal Stove

The added height allows better burning of the volatile gases and secondary combustion.

A *kitchen range* can be used for cooking as well as heating. Of course, a range warms the kitchen whenever it is used, which can be uncomfortable in the summer. Most ranges are of cast iron construction and will burn either wood or coal. Kitchen ranges are generally not used in heating the whole house.

Furnaces in many modern homes were designed for the convenience of central heating. The layout of the house may make heating with stoves virtually impossible, with many small scattered rooms sealed off from each other by walls, closets, and other heat-blocking barriers. Electrically heated homes are especially notorious for this. This type of layout is ideal for a *coal* or *multi-fuel furnace.* Coal furnaces are available that will heat any size home and are designed for using either hot air or water transfer media. In most cases the furnace is designed to replace the existing furnace using

Fig. 19—Pot Belly

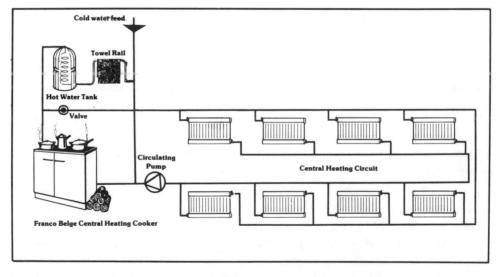

Fig. 20—A simplified installation of a Franco-Belge Model 1703 with Cooking, Hot Water and Central Heating to 8 radiators.

the pipes or vents currently in place. Maintenance of such a unit is more demanding than for a conventional oil-burner. The fire must be fed, the ash box cleaned out, and the heat exchanger cleaned. They also burn large amounts of solid fuel. Typical burning time is 12 plus hours.

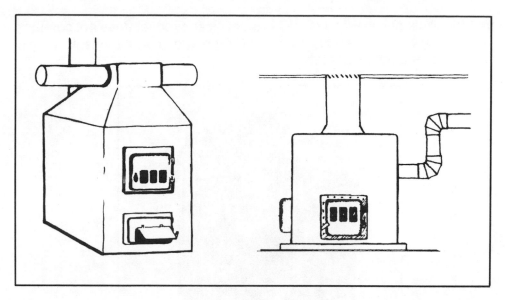

Fig. 21

In Summary

When selecting a stove the following considerations are rated in priority order:

1. Decide on what your end use will be.
2. Decide on location(s) of the stove(s).
3. Select the correct size for your application.
4. Pick a general type that best fits your use.
5. Talk to a stove dealer; tell him what you want and why.
6. Pick a stove that fits your needs and taste, not just your pocketbook.

Consider the purchasing of a stove a major investment that will pay a handsome dividend over its lifetime.

The Stove Manufacturer

When selecting a stove it is as important to examine the manufacturer as it is to examine the stove. Many stoves are sold with a five- or ten-year or even longer warranty, which is nice, but it is even nicer when the manufacturer is there five years later to honor the warranty. Coal stoves are just beginning to enjoy a period of rediscovery and popularity. This industry is no different than any industry. There are many firms just entering it that will not survive. While there is no way of accurately forecasting which ones will eventually succeed, we can be almost certain that some will not and that the likelihood is that a manufacturer whose product does not compare favorably with his competitors' from a design and/or workmanship aspect will not be among the survivors.

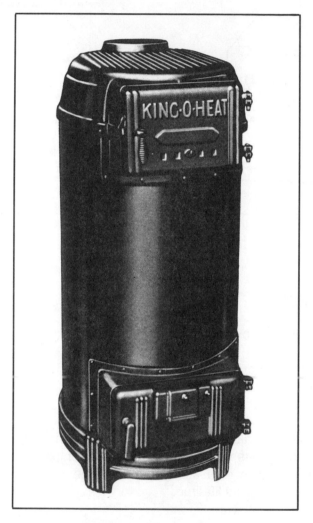

Fig. 22—King-O-Heat Coal Heater

Generally, a good retailer should be able to answer any questions about a manufacturer or at least to find the answers for you. The purchase price and installation costs of a quality coal-burning stove or furnace can range from several hundred to several thousand dollars. The answers to your questions can be worth at least the cost of postage for a letter, or a telephone call, directly to the manufacturer. Most manufacturers are conscientious and very responsive to consumer questions and concerns. This advice extends to questions and/or problems that the consumer might encounter after the purchase of a stove or furnace.

Fig. 23—Grizzly Coal Cub

Where to Buy

Coal- and wood-burning stoves can be purchased in an increasingly wide array of outlets, ranging from gas stations to hardware stores, from department stores to discount chains and from fuel dealers to stores dealing exclusively with these heating devices. The best piece of advice is to choose the stove from a dealer who allows you to examine the particular stove you have chosen before you take it home. Resist the temptation to purchase the stove from a mail order supplier or to order the device over the phone no matter what the savings appear to be. It almost goes without saying that it pays to shop in stores that have an honorable reputation and that you can expect to be in business for some time to come.

Fig. 24—Temp Coal® II

When shopping for both a stove and advice on a stove it is usually advantageous and smart to shop in retail establishments that carry a variety of manufacturers' products. Stores that carry a variety of different types of stoves as well as the necessary accessory equipment can generally answer the more difficult comparison questions. A store that is genuinely prepared to serve you and your requirements will generally travel to your home to offer advice in selecting not only the brand and model but the appropriate size and design to meet the requirements that are particular to your situation. A complete operation will offer installation services and be available to assist with the stove's operation while you are learning a new set of skills.

Chapter 7

Installation of Hand-Fired Heaters

If thine enemy hunger, feed him; if
he thirst, give hime drink: for in
so doing thou shalt heap coals
of fire on his head.

Romans, XII, 20

Coal Stoves and Furnaces

A coal stove should be installed by a competent installer who has adequate knowledge of solid fuel stove installation. Good quality and approved materials that meet Underwriters' Laboratories (UL) or other similar standards should be used, and local codes and ordinances should be observed. The stove must be at least 36 inches from combustible walls and all furnishings and there must be an 18-inch clearance between a stovepipe and any combustible material. It can be hooked up to an existing chimney, if suitable, or to a prefabricated chimney.

Many homes already have fireplaces, so the question of what to use as an alternate heat source in those instances can be easily answered. It may be possible to install a coal stove in an existing fireplace. In the absence of a fireplace, the homeowner must turn to a solid fuel stove for alternate heating.

Fig. 25

There are several factors that affect safety and should therefore be considered in the selection of an alternate heating source. These include the condition of an existing fireplace and its chimney, the available coal stove, the available means for venting, the size of the room or area to be heated, the size and shape of the stove, and the availability of air for combustion.

A new, as well as a used, coal heater should be carefully examined for cracks or other defects such as broken door hinges, warped or loose-fitting doors, faulty legs, broken or inoperative draft dampers and regulators, and broken grates. All cracks and leaks should be repaired either by welding or brazing or by using a high-temperature stove cement. Any repair work, i.e., welding or brazing, should be done by qualified persons. Only stoves having grates designed for coal should be used for coal. Grates should be in good repair.

A new coal stove should be made of steel or cast iron and should be listed by a recognized testing laboratory. It is always wiser to do business with a reputable dealer who is knowledgeable in the installation, operation, and maintenance of solid fuel heating equipment.

Some coal stoves are radiant heaters. Radiant heat usually provides greater comfort than other types of heat, such as heat from forced air systems. Radiant heat is a direct type of heat, and a family may feel

Fig. 26—Better'n Ben's Coal Stove

comfortable in an area heated to a relatively low temperature by radiant heat if the heated area is protected from drafts. Remember that an oxygen-consuming coal fire must be provided with a continuous supply of fresh air, which means that ventilation is essential, not only for proper combustion, but also for the safety of the occupants. Normal air leakage in a dwelling is usually sufficient, but may not be in a well-caulked and insulated house unless a window is left cracked open. Coal stoves and furnaces are sometimes over-fired; the stoves and their chimney connectors and chimney flues may become red and even "melt down." Over-firing, coupled with inadequate clearances and lack of constant attention, have resulted in many dangerous fires that would not otherwise have occurred.

Materials such as brick, imitation brick, slate, stone, iron, cement board or cement less than eight inches thick do not offer protection from radiant heat. The heat from coal stoves passes through these materials and will eventually dry out and ignite wood or other combustible materials on the other side. When any of the above materials are used, the sustained radiant heat from the stove is also radiated by these materials. Protective materials should dissipate rather than radiate heat. Asbestos cement board *does not* provide heat resistance. It will conduct heat to any combustible surface to which it is attached. Only proper clearance and adequate ventilation can protect against radiant heat.

Stoves and heaters are available for burning just coal. It is easy to burn wood in a coal-burning stove but it is impractical to try to burn coal in a stove designed to burn wood because of the different conditions required for the satisfactory burning of coal.

The stove or heater designed for burning coal must have a properly sized firebox, about 14 inches in diameter and at least 10 inches deep, a coal grate (not a wood grate), and both primary and secondary air supplies. Also, many coal-burning stoves are thermostatically controlled, a feature that is more important on a coal- than on a wood-burning stove so as to maintain a higher firebox temperature.

The grate supports the fuel bed so that air may be supplied to the fire and so that ashes may be removed from the fuel bed. To burn nut coal, the distance between the grate openings should be approximately three quarters of an inch; to burn pea coal, a smaller size grate opening is needed. A standard wood grate is not suitable for burning coal.

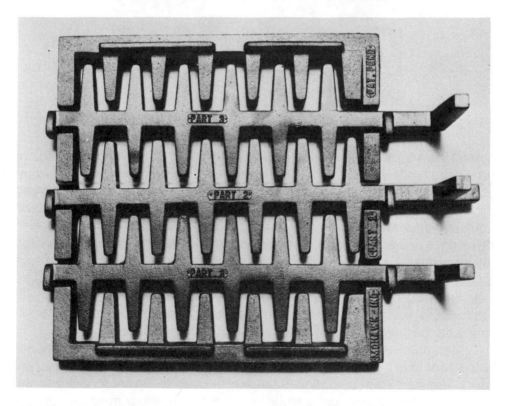

Fig. 27—Coal grates

Well-designed wood-coal combination stoves are available. All too often, though, so-called combination stoves do not burn coal efficiently. Many people who have attempted to burn coal in a combination stove have given up.

A well-designed combination stove should have removable liners to shape the firebox for efficient burning. Coal fireboxes should be shorter and deeper than wood fireboxes. There should be space for circulation of air between the coal grate and the bottom of the stove or ash pan, and a way to shake ashes from the fire. The stove should have an ash pan for convenient ash removal; a properly designed coal grate that can be replaced with a wood grate if wood is to be burned; and a slightly larger draft opening for burning coal.

Coal is becoming a favorite alternative fuel. Attracted by the prospect of saving energy and money, thousands of homeowners all across the nation are installing coal stoves. Unfortunately, many of these stoves are being improperly installed or carelessly operated so stove fires are becoming more and more common. Coal stoves are not inherently dangerous; most of these fires could have been easily prevented by common safety measures. Safety is not an exciting subject, but it deserves your attention as if your life depended on it, for it well may. When installing a stove, you must use more than your common sense as your guide. Trial and error is no way to learn about safety. You must be right the first time.

Remember that above all else, a coal stove should be *safe*.

The clearance and installation guidelines presented here are based on the standards of the National Fire Protection Association and represent many years of experience. They provide reasonable, but not extreme, margins of safety.

However, some states and municipalities have specific regulations that apply to the installation of coal stoves, and these may be different from the ones presented here. Check with your local building inspector, fire department, or state energy office to see if a permit is needed or if more specific regulations apply to your situation. To comply with the law, be sure to use the most stringent of the installation guidelines.

Insurance agents should also be notified to ensure coverage.

Coal Stove Clearances

The National Fire Protection Administration (NFPA) standards recommend a 36-inch clearance between a coal stove and any combustible material and an 18-inch clearance between a stovepipe and any combustible material.

Wallboard or sheetrock *is* considered a combustible because it conducts heat to the combustible behind it.

Remember that these clearances are for all combustible materials. This eliminates some dangerous but common practices such as stacking wood or paper next to a stove or moving a sofa "temporarily" closer. It may be difficult to install a stove where it is convenient, aesthetically pleasing, and safe, but don't compromise. Safety is your priority.

At one time separate standards were specified for stoves with sheet metal cabinets (also called circulating stoves), but the National Fire Protection Association now recommends that all stoves be treated the same.

TABLE 7
MINIMUM CLEARANCES

Type of Protection	Stove Sides	Stovepipe
Unprotected	36"	18"
¼" asbestos millboard	36"	18"
¼" asbestos millboard spaced out 1"	18"	12"
28-gauge sheet metal on 14" asbestos millboard	18"	12"
28-gauge sheet metal spaced out 1"	12"	9"
28-gauge sheet metal on ⅛" asbestos millboard spaced out	12"	9"
22-gauge sheet metal on 1" mineral fiber bats reinforced with wire mesh or equivalent	12"	12"

The use of asbestos millboard in wall and floor protection is a controversial issue because of the health hazard of asbestos fibers in the manufacture and in the preparation and handling of the millboard for use. The National Fire Protection Association is currently initiating the process of removing asbestos as a standard protection for reduced clearances. We would strongly encourage use of an alternative protection whenever one is available. However, if you must use asbestos millboard, use it cautiously. We recommend painting the asbestos to help keep the fibers from coming loose. If the board must be cut, do not inhale the dust. Do the work outdoors, using a breathing mask.

Floor Protection for Coal Heaters

Safe floor clearances are substantially less than those for walls because the heat radiated from the bottom of a stove is generally less than from either the sides or the top. During a fire, ashes fall to the bottom of a stove. This has an insulating effect, resisting the flow of heat downward. This process may be started in a new stove by placing a layer of sand in the bottom.

Clearances for proper floor protection are classified according to leg length, and are listed below.

TABLE 8

Leg length	Protection needed
18 inches or more	24-gauge layer of sheet metal
6 to 18 inches	24-gauge layer of sheet metal over ¼-inch layer of asbestos millboard
6 inches or less	4 inches of hollow masonry laid to provide air circulation through the masonry layer covered by a sheet of 24 gauge sheet metal.

Floor protection materials can be covered with attractive materials such as stone, brick, or tile, as long as they are non-combustible, but do not use these materials instead of an asbestos sheet or metal plate.

Falling embers and sparks present an additional safety problem that is often ignored. Avoid this potential problem by extending the floor protection 18 inches from the front of the stove and 12 inches around the sides and back. This affords a reasonable amount of protection, but you should still take care when loading and tending the stove. Make sure that ashes and hot coals fall only on the protected area.

Wall Protection

The recommended clearances can be reduced considerably if combustible walls and ceilings are protected with asbestos millboard or 28-gauge sheet metal spaced out one inch from the combustible wall. The spacers should be constructed from a non-combustible material. Provide a one-inch air gap at the bottom of the asbestos millboard or metal panel. Air circulating behind the panel will cool the panel and the wall.

Asbestos millboard is a different material from asbestos cement board. Asbestos cement board (transite) is designed as a flame barrier. It provides little in terms of heat resistance—it will conduct heat to any combustible surface to which it is attached. Brick and stone also provide little or no protection for a combustible wall because they are good conductors of heat. To be effective, bricks must be spaced out an inch from the wall with air gaps at the top and bottom. This can be accomplished by using half bricks in the top and bottom rows.

Drywall (gypsum wall board) over studs is considered a combustible wall. Heat is transmitted directly through the dry wall to the studs.

Installing a Coal Stove in an Existing Fireplace

This is something that should not be done, but if you must, here are some things of concern. First, a fireplace chimney is sized for the large air flow of a wide open fire. Airtight stoves do not use as much air and therefore don't need such a large flue. An oversized flue means a much larger surface area to cool the gases, i.e., backdraft.

If you still decide to use this type of installation, there are three ways to accomplish it. The pipe can: (1) be vented into the fireplace only; (2) be vented into the fireplace and part way up the chimney; or (3) enter the chimney by a new hole above the mantle.

In all cases, all openings into the fireplace should be thoroughly sealed.

All of these clearances can be reduced by using a heat shield of approved materials and suspending it one inch from the wall using non-combustible spacers. Ceramic fence insulators make good non-combustible spacers.

Installing a coal stove or furnace in accordance with these specifications could prevent a serious fire.

Recent testing by the NFPA on safe clearances for coal stoves has come up with some interesting results. Because each stove has a unique radiation pattern, the effects on nearby combustibles can be very different for various stoves.

In fact, the safe distance from combustibles for one stove model was determined to be 52 inches; the 36-inch figure should, then, be used as a guide, with a larger stove possibly requiring a greater clearance.

The only way to be absolutely sure is to buy a coal stove that has been listed by a testing laboratory (such as Underwriters' Laboratories) and follow its recommendations.

If your coal heater is not listed, check the wall surfaces regularly to find out if they are hot to the touch. If the walls become too hot to touch, it is a wise idea to install some type of protection, as described earlier.

Be especially careful if a coal stove is installed near an exterior *insulated* wall. Because the wall is insulated, it is unable to dissipate the heat as rapidly as an uninsulated wall. Therefore, clearances may have to be greater or protection may have to be provided.

The most important thing to remember when installing a protective shield is to allow for *air circulation* behind the shield. This is accomplished by spacing the shield away from the wall and leaving at least a one-inch opening at the top and bottom of the shield.

There are certain procedures you should follow to make sure the installation is safe. The suggestions included herein are minimal safety procedures. You should contact your local authority (fire department or building inspector) for any special requirements or building restrictions.

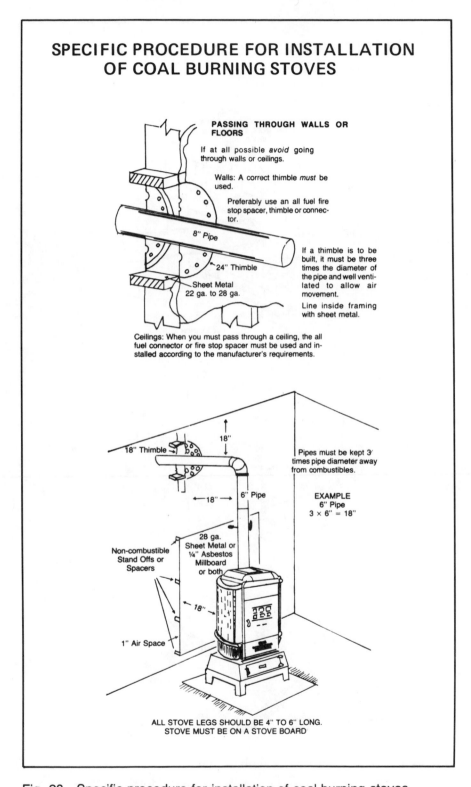

SPECIFIC PROCEDURE FOR INSTALLATION OF COAL BURNING STOVES

PASSING THROUGH WALLS OR FLOORS

If at all possible *avoid* going through walls or ceilings.

Walls: A correct thimble *must* be used.

Preferably use an all fuel fire stop spacer, thimble or connector.

8" Pipe

24" Thimble

Sheet Metal 22 ga. to 28 ga.

If a thimble is to be built, it must be three times the diameter of the pipe and well ventilated to allow air movement.

Line inside framing with sheet metal.

Ceilings: When you must pass through a ceiling, the all fuel connector or fire stop spacer must be used and installed according to the manufacturer's requirements.

18"

18" Thimble

18"   6" Pipe

Pipes must be kept 3 times pipe diameter away from combustibles.

EXAMPLE
6" Pipe
3 × 6" = 18"

28 ga. Sheet Metal or ¼" Asbestos Millboard or both

Non-combustible Stand Offs or Spacers

18"

1" Air Space

ALL STOVE LEGS SHOULD BE 4" TO 6" LONG.
STOVE MUST BE ON A STOVE BOARD

Fig. 28—Specific procedure for installation of coal-burning stoves

SPECIFIC PROCEDURE FOR INSTALLATION OF COAL BURNING STOVES

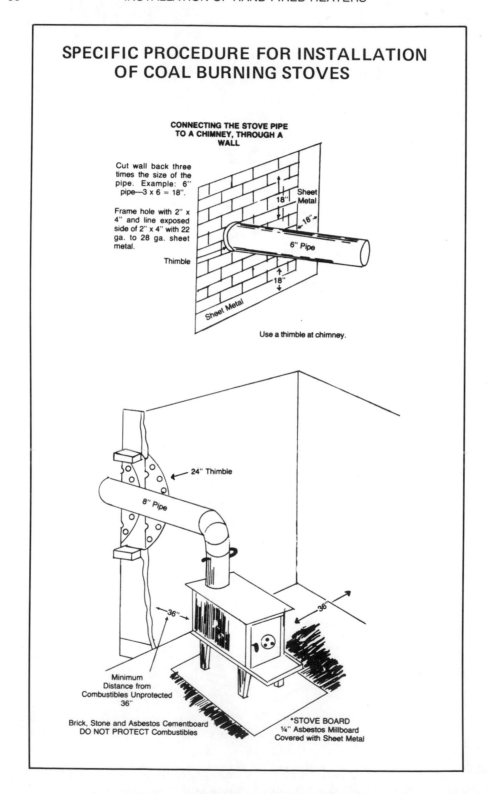

CONNECTING THE STOVE PIPE TO A CHIMNEY, THROUGH A WALL

Cut wall back three times the size of the pipe. Example: 6" pipe—3 x 6 = 18".

Frame hole with 2" x 4" and line exposed side of 2" x 4" with 22 ga. to 28 ga. sheet metal.

Thimble

Sheet Metal

18"

18"

6" Pipe

18"

Sheet Metal

Use a thimble at chimney.

24" Thimble

8" Pipe

36"

36"

Minimum Distance from Combustibles Unprotected 36"

Brick, Stone and Asbestos Cementboard DO NOT PROTECT Combustibles

*STOVE BOARD
¼" Asbestos Millboard Covered with Sheet Metal

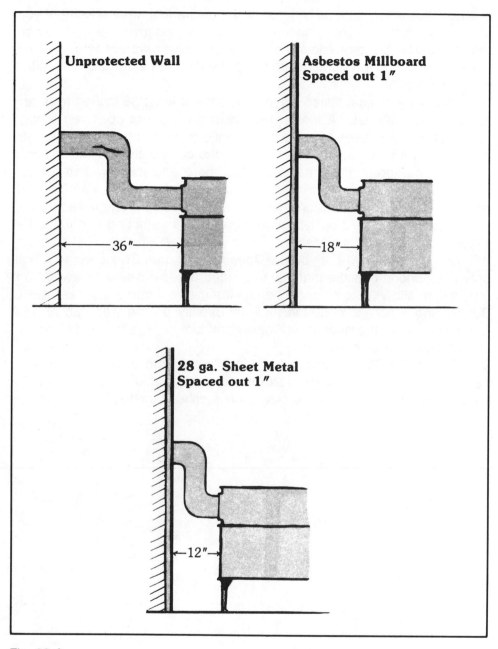

Fig. 28 A

Siting a Coal Stove

A coal heater must be a certain distance away from combustible materials to prevent fires from radiated heat. If your stove is listed by a national testing laboratory (such as UL), follow the clearances they specify. If your stove is not listed, diligently adhere to the clearances given here. Be aware that the sides, bottom, and stovepipe attain quite different temperatures while the stove is in operation, so there are different clearances specified for each of them.

The sides of an unlisted stove should be at least 36 inches from any combustible materials. A foot or two clearance will not do. Even though very high temperatures are needed to ignite most combustible materials, high temperatures over time can change the composition of the material by slowly darkening it so that it accepts more and more radiated heat. Finally it may begin to smolder and burn at temperatures as low as 200–250 degrees F. These temperatures are easily reached by an unprotected wall exposed to a stove without adequate clearance, so strict adherence to the clearance standards is advised.

An unlisted stove may be positioned closer than 36 inches to a wall only if a noncombustible material is spaced at least one inch away from the wall, to allow air to circulate behind the material and carry heat away. But placing a non-combustible material directly on the wall has no real protective value; the material will easily conduct heat to the wall behind it, creating dangerous conditions.

Asbestos millboard spaced one inch away from a wall will allow an 18-inch stovewall clearance. If 28-gauge sheet metal is installed in the same manner, the acceptance clearance is only 12 inches.

Chapter 8

Chimneys and Flues

Golden lads and girls all must,
As chimney-sweepers, come to dust.
Shakespeare

One of the most important aspects of heating with coal is the chimney. This chapter deals with the standards recommended by the National Fire Protection Association, regional building associations, state authorities, and local building inspectors. All of these standards provide for height, width, wall thickness, choice of materials, location of the chimney in the house, chimney size, and other concerns to the coal-burning homeowner.

Proper construction of chimneys is essential for safe, efficient operation of any hand-fired coal heater. Therefore, it is recommended that they be designed and built by persons experienced in that type of work. The home-owner should have a working knowledge of chimney construction so that he can assist in the designing, follow the work closely, and properly inspect and maintain the completed unit.

Chimneys for coal heaters have changed little since devices to conduct smoke away from the living space first appeared in medieval Europe. In the early 1900's, fire-resistant clay flue liners first appeared in smokestacks in the United States, and the most recent developments in chimney tech-nology began to appear in new houses during the 1950's with insulated steel and triple-wall prefabricated chimneys.

Chimneys present an uncertain danger. Most of them are concealed or inaccessible, yet they play a critical role in the overall safety of your coal-burning system. Be sure that your chimney will safely do the job.

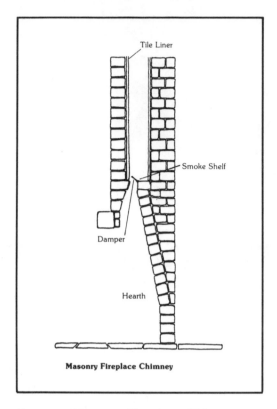

Masonry Fireplace Chimney

Fig. 29—Masonry Fireplace Chimney

Chimneys have two basic functions. They create a draft to bring air to the fire, and they carry hot gases safely away. To accomplish this, two types of chimneys are commonly used: prefabricated metal and masonry. According to tests conducted by the National Bureau of Standards, they will produce equal drafts under similar conditions.

Chimney Construction

All fuel-burning equipment such as the coal stove requires some type of chimney (Fig. 30). The chimney must be designed and built so that it produces sufficient draft to supply an adequate quantity of fresh air to the fire and to expel smoke and gases emitted by the fire or equipment.

A chimney located entirely inside a building has better draft than an exterior chimney, because the masonry retains heat longer when protected from cold outside air.

Masonry chimneys are more costly to construct than prefabricated metal chimneys, but they have other benefits that may outweigh their high cost. For example, a chimney built inside the house has the capacity to

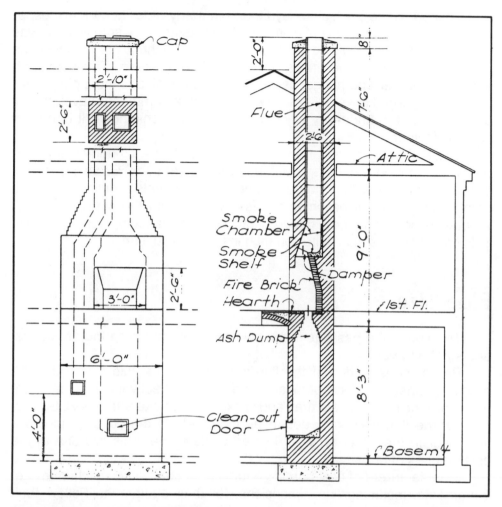

Fig. 30—Diagram of an entire chimney such as is commonly built to serve the house-heating unit and one fireplace.

store excess heat and temper the interior environment. That is one of the reasons why in colonial New England it was common practice to build houses with large center chimneys. These inside masonry chimneys increase the net efficiency of the heating system by up to 15 percent due to the "leaking" of heat from the chimney into the house. In addition to being more attractive, masonry chimneys tend to be more durable because they are more resistant to corrosion.

The masonry chimney will hold the heat much like the cast iron or steel in the stove. It may keep the pipes from freezing if you don't get home. It will also draw cool air into the house during the summer. A brick chimney can be custom-made to include a smoke chamber or water-heating coils for preheating domestic hot water and can be an attractive addition to your home.

Unlined single-brick chimneys found in many older homes are especially hazardous. This type of chimney often was not very safe when it was built and certainly should be suspect now. Mortar in the joints probably has broken down and some bricks or tiles may be cracked. The combined action of weather and hot gases causes these conditions most often near the chimney top. However, cracks and openings commonly develop well below the roof in tinder-dry attics. Do not use the chimney if it does not have a tile liner.

Cracks in chimneys can be located by building a smudge fire in the bottom, then covering the top with a board or wet sack. Escaping smoke should readily reveal the chimney's condition. All defects should be repaired before use even if it becomes necessary to rebuild the whole chimney. Woodwork should not be in direct contact with the masonry of any chimney. This condition was also quite common in old construction.

Flue Size

The flue is the passage in the chimney through which the air, gases, and smoke travel.

Proper construction of the flue is important. Its size (area), height, shape, tightness, and smoothness determine the effectiveness of the chimney in producing adequate draft and in expelling smoke and gases. Soundness of the flue walls may determine the safety of the building should a fire occur in the chimney. Over-heated or defective flues are one of the chief causes of house fires.

Manufacturers of coal-burning equipment usually specify chimney requirements, including flue dimensions, for their equipment. Follow their recommendations.

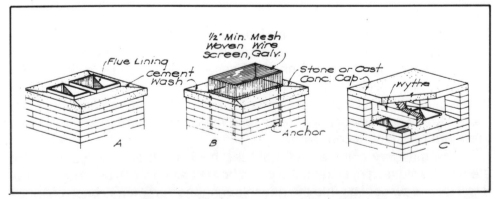

Fig. 31—Top construction of chimneys. A, Good method of finishing top of chimney; flue lining extends 4 inches above cap. B, Spark arrester or bird screen. C, Hood to keep out rain.

Unfortunately, some existing chimneys used with coal stoves are too large, have walls too thin, are without flue linings, and lack proper clearances from combustible structural members of buildings. Chimneys capable of providing the required amount of draft specified by the coal stove manufacturer must be provided.

Height

A chimney should extend at least three feet above flat roofs and at least two feet above a roof ridge or raised part of a roof within ten feet of the chimney. A hood should be provided if a chimney cannot be built high enough above a ridge to prevent trouble from eddies caused by wind being deflected from the roof. The open ends of the hood should be parallel to the ridge.

Low-cost metal-pipe extensions are sometimes used to increase flue height, but they are not as durable or as attractive as terra cotta chimney pots or extensions. Metal extensions must be securely anchored against the wind and must have the same cross-sectional area as the flue. They are available with a metal cowl or top that turns with the wind to prevent air from blowing down the flue.

Support

The chimney is usually the heaviest part of a building, and it must rest on a solid foundation to prevent differential settlement in the building.

Concrete footings are recommended. They must be designed to distribute the load over an area wide enough to avoid exceeding the safe load-bearing capacity of the soil. They should extend at least six inches beyond the chimney on all sides and should be eight inches thick for one-story houses and twelve inches thick for two-story houses having basements.

If there is no basement, pour the footings for an exterior chimney on solid ground below the frostline.

If the house wall is of solid masonry at least twelve inches thick, the chimney can be built integrally with the wall and, instead of being carried down to the ground, it can be offset from the wall enough to provide flue space by corbelling. The offset should not extend more than six inches from the face of the wall—each course projecting not more than one inch—and should be not less than twelve inches high.

Chimneys in frame buildings should be built from the ground up, or they can rest on the building foundation or basement walls if the walls are of solid masonry 12 inches thick and have adequate footings.

Flue Lining

Chimneys are sometimes built without flue linings to reduce cost, but those with lined flues are safer and more efficient.

Lined flues are definitely recommended for brick chimneys. When the flue is not lined, mortar and bricks directly exposed to the action of flue gases disintegrate. This disintegration plus that caused by temperature changes can open cracks in the masonry, which will reduce the draft and increase the fire hazard.

Flue lining must withstand rapid fluctuations in temperature and the action of flue gases. Therefore, it should be made of vitrified fire clay at least five-eighths of an inch thick.

Both rectangular- and round-shaped linings are available. Rectangular lining is better adapted to brick construction, but round lining is more efficient.

Each length of lining should be placed in position—set in cement mortar with the joint struck smooth on the inside—and then the brick laid around it. If the lining is slipped down after several courses of brick have been laid, the joints cannot be filled and leakage will occur. In masonry chimneys with walls less than eight inches thick, there should be space between the lining and the chimney walls. This space should not be filled with mortar. Use only enough mortar to make good joints and to hold the lining in position.

Unless it rests on solid masonry at the bottom of the flue, the lower section of lining must be supported on at least three sides of brick courses projecting to the inside surface of the lining. This lining should extend to a point at least eight inches under the smoke pipe thimble or flue ring (see Fig. 35).

Flues should be as nearly vertical as possible. If a change in direction is necessary, the angle should never exceed 45° (Fig. 32). An angle of 30° or less is better, because sharp turns set up eddies which affect the motion of smoke and gases. Where a flue does change directions the lining joints should be made tight by mitering or cutting equally the ends of the adjoining sections. Cut the lining before it is built into the chimney; if cut after, it may break and fall out of place. To cut the lining, stuff a sack of damp sand into it and then tap a sharp chisel with a light hammer along the desired line of cut.

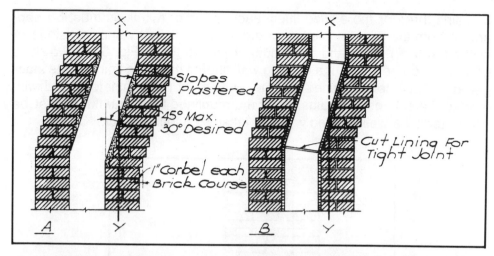

Fig. 32—Offset in a chimney. For structural safety the amount of offset must be limited so that the center line, XY, of the upper flue will not fall beyond the center of the wall of the lower flue. A, Start the offset of the left wall of an unlined flue two brick courses higher than the right wall so that the area of the sloping section will not be reduced after plastering. B, Method of cutting lining to make a tight joint.

When laying lining and brick, draw a tight-fitting bag of straw up the flue as the work progresses to catch material that might fall and block the flue.

Walls

Walls of chimneys with lined flues and not more than 30 feet high should be at least four inches thick if made of brick or reinforced concrete and at least twelve inches thick if made of stone.

Flue lining is recommended, especially for brick chimneys, but it can be omitted if the chimney walls are made of reinforced concrete at least six inches thick or of unreinforced concrete or brick at least eight inches thick.

A minimum thickness of eight inches is recommended for the outside wall of a chimney exposed to the weather.

Brick chimneys that extend up through the roof may sway enough in heavy winds to open up mortar joints at the roof line. Openings to the flue at that point are dangerous, because sparks from the flue may start fires in the woodwork or roofing. A good practice is to make the upper walls eight inches thick by starting to offset the bricks at least six inches below the underside of roof joists or rafters (Fig. 32).

Chimneys may contain more than one flue. Building codes generally require a separate flue for each fireplace, furnace, or boiler. If a chimney

contains three or more lined flues, each group of two flues must be separated from the other single flue or group of two flues by brick divisions or
wythes at least three and three-quarter inches thick (Fig. 33). Two flues
grouped together without a dividing wall should have the lining joints staggered at least seven inches and the joints must be completely filled with
mortar. If a chimney contains two or more unlined flues, the flues must be
separated by a well-bonded wythe at least eight inches thick.

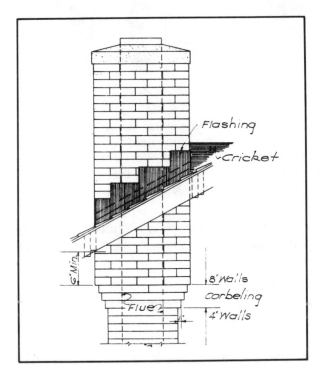

Fig. 33—Corbelling of chimney to provide 8-inch
walls for the section exposed to the weather.

Mortar

Brickwork around chimney flues and fireplaces should be laid with
cement mortar; it is more resistant to the action of heat and flue gases than
lime mortar.

A good mortar to use in setting flue linings and all chimney masonry,
except firebrick, consists of one part Portland cement, one part hydrated
lime (or slaked-lime putty), and six parts clean sand, measured by volume.

Firebrick should be laid with fire clay.

Soot Pocket and Cleanout

A soot pocket and cleanout are recommended for each flue (Fig. 35).

Deep soot pockets permit the accumulation of an excessive amount of soot, which may take fire. Therefore, the pocket should be only deep enough to permit installation of a cleanout door below the smokepipe connection. Fill the lower part of the chimney—from the bottom of the soot pocket to the base of the chimney—with solid masonry.

The cleanout door should be made of cast iron and should fit snugly and be kept tightly closed to keep air out.

A cleanout should serve only one flue. If two or more flues are connected to the same cleanout, air drawn from one to another will affect the draft in all.

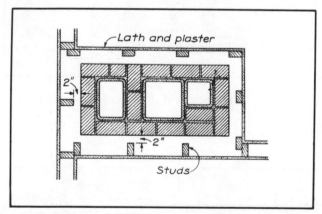

Fig. 34—Plan of chimney showing proper arrangement of three flues. Bond division wall with sidewalls by staggering the joints of successive courses. Wood framing should be at least 2 inches from brickwork.

Chimney Smokepipe Connection

No range, stove, fireplace, or other equipment should be connected to the flue for the central heating unit. In fact, as previously indicated, each unit should be connected to a separate flue, because if there are two or more connections to the same flue, fires may occur from sparks passing into one flue opening and out through another.

Smokepipes from furnaces, stoves, or other equipment must be correctly installed and connected to the chimney for safe operation.

A smokepipe should enter the chimney horizontally and should not extend into the flue (Fig. 36). The hole in the chimney wall should be lined with fire clay, or metal thimbles should be tightly built into the masonry. (Metal thimbles or flue rings are available in diameters of six, seven, eight, ten, and twelves inches.) To make an airtight connection where the pipe enters the wall, install a closely fitting collar and apply boiler putty, good cement mortar, or stiff clay.

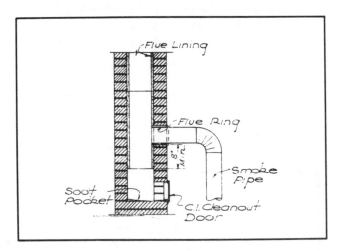

Fig. 35—Soot pocket and cleanout for a chimney flue.

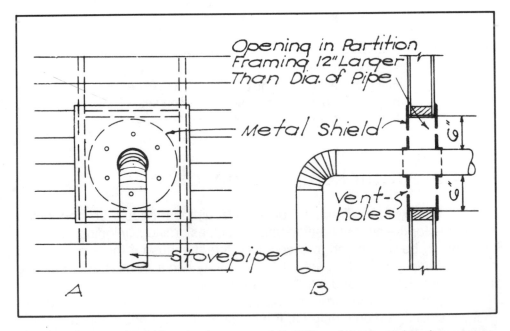

Fig. 36—One method of protecting a wood partition when a smoke pipe passes through it. A, Elevation of protection around the pipe. B, Sectional view.

A smokepipe should never be closer than nine inches to woodwork or other combustible material. If it is less than eighteen inches from woodwork or other combustible material, cover at least the half of the pipe nearest the woodwork with fire-resistant material. Commercial fireproof pipe covering is available.

If a smokepipe must pass through a wood partition, the woodwork must be protected. Either cut an opening in the partition and insert a galvanized-iron, double-wall ventilating shield at least twelve inches larger than the pipe (Fig. 37) or install at least four inches of brickwork or other incombustible material around the pipe.

Smokepipes should never pass through floors, closets, or concealed spaces or enter the chimney in the attic.

Smoke Test

Every flue should be tested as follows before being used and preferably before the chimney has been furred, plastered, or otherwise enclosed.

Build a paper, straw, wood, or tar-paper fire at the base of the flue. When the smoke rises in a dense column, tightly block the outlet at the top of the chimney with a wet blanket. Smoke that escapes through the masonry indicates the location of leaks.

This test may show bad leaks into adjoining flues, through the walls, or between the lining and the wall. Correct defects before the chimney is used. Since such defects may be hard to correct, you should check the initial construction carefully.

Each summer when they are not in use, smokepipes should be taken down, cleaned, wrapped in paper, and stored in a dry place.

When not in use, smokepipe holes should be closed with tight-fitting metal flue stops. Do not use papered tin. If a pipe hole is to be abandoned, fill it with bricks laid in good mortar. Such stopping can be readily removed if necessary.

Insulation

No wood should be in contact with the chimney. Leave a two-inch space (Fig. 34) between the chimney walls and all wooden beams or joists (unless the walls are of solid masonry eight inches thick, in which case the framing can be within one-half inch of the chimney masonry).

Fill the space between wall and floor framing with porous, non-metallic, incombustible material, such as loose cinders (Fig. 37). Do not use brickwork, mortar, or concrete. Place the filling before the floor is laid, because it not only forms a firestop but also prevents the accumulation of shavings or other combustible material.

Flooring and subflooring can be laid within three-fourths-inch of the masonry. Wood studding, furring, or lathing should be set back at least two inches from chimney walls. (Plaster can be applied directly to the masonry or to metal lath laid over the masonry, but this is not recommended because settlement of the chimney may crack the plaster.) A coat of cement plaster should be applied to chimney walls that will be encased by wood partition or other combustible construction.

If baseboards are fastened to plaster that is in direct contact with the chimney wall, install a layer of fireproof material, such as asbestos, at least one-eighth-inch thick between the baseboard and the plaster (Fig. 37).

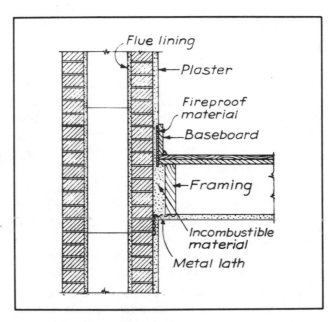

Fig. 37—Method of insulating wood floor joists and baseboard at a chimney.

Connection with Roof

Where the chimney passes through the roof, provide a two-inch clearance between the wood framing and the masonry for fire protection and to permit expansion due to temperature changes, settlement, and slight movement during heavy winds.

Chimneys must be flashed and counterflashed to make the junction with the roof watertight (Figs. 39 and 40). When the chimney is located on the slope of a roof, a cricket (j, Fig. 38) is built as shown in Fig. 40 high enough to shed water around the chimney. Corrosion-resistant metal, such as copper, zinc, or lead, should be used for flashing. Galvanized or tinned sheet steel requires occasional painting.

Top Construction

Fig. 31,A shows a good method of finishing the top of the chimney. The flue lining extends at least four inches above the cap or top course of brick and is surrounded by at least two inches of cement mortar. The mortar is finished with a straight or concave slope to direct air currents upward at the top of the flue and to drain water from the top of the chimney.

Hoods (Fig. 31,C) are used to keep rain out of chimneys and to prevent downdraft due to nearby buildings, trees, or other objects. Common types are the arched brick hood and the flat-stone or cast-concrete cap. If the hood covers more than one flue, it should be divided by wythes so that

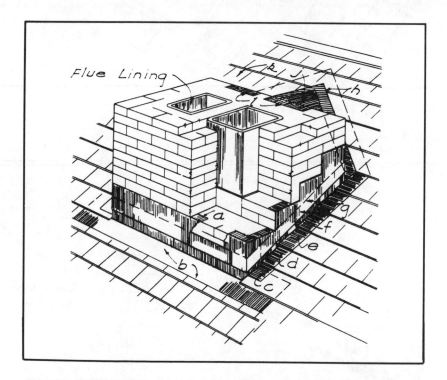

Fig. 38—Flashing at a chimney located on the slope of a roof. Sheet metal (h), over cricket (j), extends at least 4 inches under the shingles (k), and is counterflashed at l in joint. Base flashings (b,c,d, and e) and cap flashings (a,f, and g) lap over the base flashings to provide watertight construction. A full bed of mortar should be provided where cap flashing is inserted in joints.

each flue has a separate section. The area of the hood opening for each flue must be larger than the area of the flue.

Spark arresters (Fig. 31,B) are recommended when burning fuels that emit sparks, such as sawdust, or when burning paper or other trash. They may be required when chimneys are on or near combustible roofs, woodland, lumber, or other combustible material. They are not recommended when burning soft coal, because they may become plugged with soot.

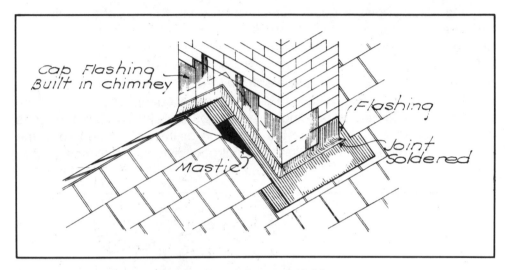

Fig. 39—Flashing at a chimney located on a roof ridge.

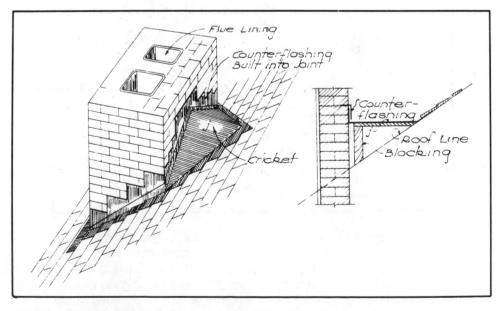

Fig. 40—Construction of a cricket (j, fig. 38) behind a chimney.

Spark arresters do not entirely eliminate the discharge of sparks, but if properly built and installed they greatly reduce the hazard. They should be made of rust-resistant material and should have screen openings not larger than five-eighths of an inch nor smaller than five-sixteenths of an inch. (Commercially made screens that generally last for several years are available.) They should completely enclose the flue discharge area and must be securely fastened to the top of the chimney. They must be kept adjusted in position and they should be replaced when the screen openings are worn larger than normal size.

Chimney Retrofit for Coal

Most of us won't be installing a new chimney for our stove; we will be using an old one. Perhaps we can be sure about the safety of a newly installed chimney because we can check it during construction, but we have no such luxury with an old chimney. We have no way of knowing the quality of the initial job, and it is difficult to check its durability over time. The best thing to do is to get the opinion of an expert.

Ask an official of the local fire department to make a chimney inspection. They generally welcome the change to practice some "preventive medicine." A competent mason may be able to point out defects, and the local building inspector can advise you about compliance with building codes. There are also some things you can check for.

First, look for a tile liner. Not all old chimneys have one, but their proven safety value makes them essential. These liners are made of material like a clay flower pot, only harder. Look up or down into the chimney flue with a flashlight; the flue should look smooth and uniform. Scrape its surface and you will find that it is much harder than brick. If you don't have a liner, one must be installed in the old chimney or the chimney must be completely rebuilt. The flue tiles are brittle, and they sometimes crack. You must be sure that the liner is in good condition. Check for loose mortar by scraping between the bricks. If the mortar crumbles easily, you will need to have the bricks repointed, the old mortar scraped out, the brick removed, new mortar installed, then the brick replaced.

Here are some "don'ts" to be aware of:

1. Don't use a masonry chimney that is supported by brackets or shelves. The brackets may weaken and the chimney may fall while in use.

2. Don't use a chimney that is used to support the framing of the house. Heat conducted to the framing may eventually start a fire. For interior chimneys a full two inches of clearance between the chimney and any combustible materials is required. Zero clearance is acceptable if the chimney is eight inches thick at that point. An exterior chimney may have one side in contact with exterior sheating.

3. Don't connect more than one appliance to a flue. Backdrafts may force carbon monoxide into your home. A fireplace flue may be used for a stove only if the fireplace opening is closed off by a non-combustible material.

4. Don't use snap-in covers to close unused smokepipe inlets. A chimney fire can easily pop them out. Fill the hole with masonry or tile built up to the same thickness as the chimney.

5. Don't use a chimney whose flue isn't at least the same size as the stovepipe for the coal stove. This will help prevent backdrafts.

Correct installation and maintenance of the chimney flue, chimney connectors (stovepipes), and joints are essential for the safe operation of a solid fuel heating system. The modern airtight, high-efficiency coal stoves only waste about ten percent of the heat up your flue. A wood stove by comparison wastes upwards of forty percent or more of the heat it produces. This is desirable and necessary because of the creosote problems inherent in wood burning. The large amount of heat sent up the flue ensures a good draft, in most cases. With a high-efficiency coal stove the stack temperature (temperature of gases exiting through the stovepipe) is several hundred degrees cooler, and the resultant draft is therefore less. For this reason an interior chimney flue measurement of greater than 7 × 7 may leave too large a volume of air to be properly heated by the small amount of wasted heat in your coal unit. Many of the standard flue sizes (8 × 12, 7 × 11, 14 × 14) are just too large to provide sufficient draft and should be lined with a six-inch liner. This cuts down on air volume and increases draft. It is highly recommended that you have your stove and chimney installed by a competent individual who can check your chimney draft and offer advice about lining your existing chimney to insure proper and safe operation of your new hand-fired coal heater.

New Chimney Liner

Masonry chimneys used with coal stoves with a large cross-sectional area, especially if on the outside of the house, may cool flue gases excessively. They may also be large enough to permit cool air to flow down the chimney, cooling flue gases, a problem that may be solved with a cap on the top of the chimney that reduces the size of the hole at the top. On other chimneys, an old cap may restrict flow excessively, in which case it should be removed or replaced. Since existing chimneys were usually built for a fireplace or furnace, it may be necessary to run a stainless steel flue inside the old flue liner, and insulate around it to make it suitable for your stove.

Lining a fireplace chimney differs from the ordinary lining in two ways: First, the hand-fired heater is attached directly to the liner without the use

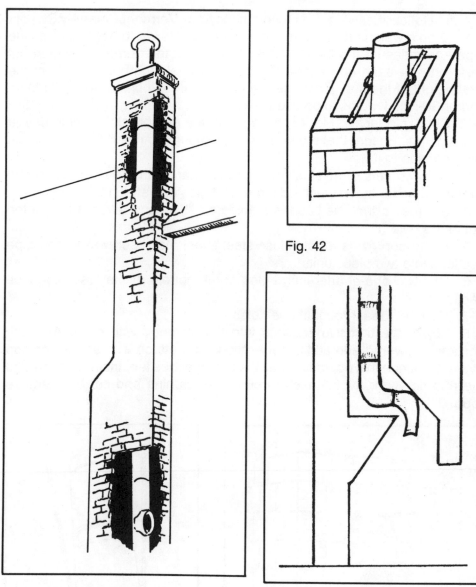

Fig. 42

Fig. 41—A flue liner that makes a faulty chimney serviceable.

Fig. 43

of a tee. Therefore, the liner must be anchored at the top. This is accomplished by attaching two "U" bolts to the last section of pipe and running two rods transversely through them.

The chimney is then capped in the usual way.

Secondly, the heating unit is connected with the use of one or more elbows, which brings the liner down through the smoke chamber for easy attachment. The space around the liner is then filled with insulation or a sheet metal plate.

Just follow these steps as recommended by Vermont Chimney Sweep:

1. Drop a weighted rope into the chimney and lower to base. Extract the rope and measure the length to be lined. Count the number of clean-out tees (one for each appliance). Subtract the total length of tees from the total height of the flue. Compute the number of 12-inch, 18-inch, 24-inch, and/or 36-inch lengths to be used.

2. Choose a location for each clean-out tee and break into chimney at these points. Ensure that the hole is large enough to insert tee and attach lowered pipe to it.

3. Drill two sets of holes slightly further apart than the diameter of the liner to be used. Insert steel rods through holes and attach "S" hooks.

4. Set tee so that the base is cradled on "S's" and take-off is pointed out of the hole.

5. Using pop rivets (or stainless steel sheet metal screws) attach pipe sections into two-piece units.

6. Drill two one-quarter-inch holes at the bottom of the first two-piece section.

7. Run the rope through these holes.

8. Lower the first two sections into the chimney with rope. As each section is lowered add additional sections and attach with screws or pop rivets. Continue this operation until the liner reaches tee. Insert section into tee and screw or rivet. Reach through tee opening and cut and remove rope.

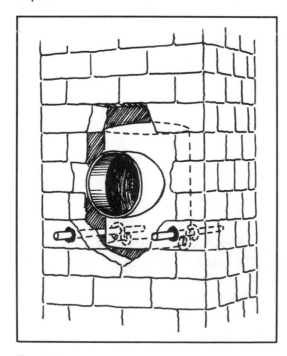

Fig. 44

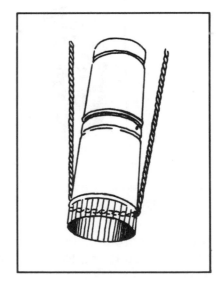

Fig. 45

9. If more than one tee is to be used, repeat steps 6 through 8, attaching second tee to top of last section of liner.

10. Mortar tee(s) into chimney. Take-off which is mortared into walls will support weight.

11. Top chimney with sheet metal plate that overlaps edge of chimney by two inches and fold down. Cut a hole into the center of the plate so that the top of the liner sticks through. This will hold and center the liner.

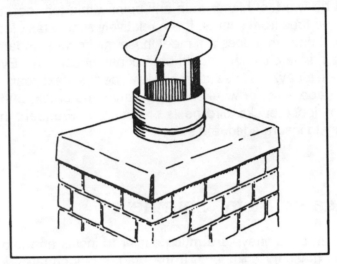

Fig. 46

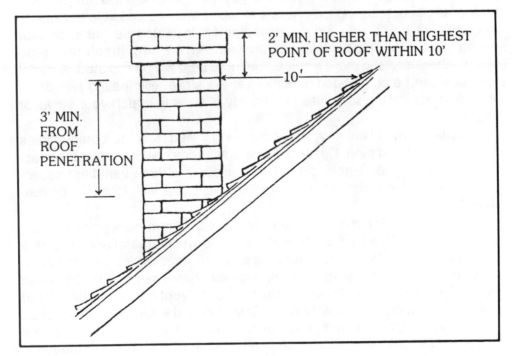

Fig. 47

12. Caulk plate to chimney, sealing all edges against water leakage. Paint sheet metal with heat-proof paint. Install rain cap and screw into place. Use screws so that cap can be removed for cleaning of liner by your chimney sweep.

You now have a SAFE chimney for a hand-fired coal stove.

The final area for inspection is chimney height. A masonry chimney should extend at least three feet above flat roofs. On pitched roofs, chimneys should be two feet higher than any point within ten feet, to prevent downdrafts and fires from sparks. It is a relatively simple task for a mason to extend a chimney that does not meet these minimum standards.

Keeping a safe chimney should be a constant concern. Even if your chimney is safe today, this does not mean it will be safe next year. Chimneys get tired with age—mortar weakens, they settle and crack, and they can be blocked by loose bricks, bird nests or beehives. Frequent inspections are necessary to insure adequate safety.

Prefabricated Metal Chimneys

Prefabricated chimneys are much easier to install and allow greater flexibility than masonry chimneys. If the draft is inadequate, you merely add another section of pipe. If you find your flue gases are too hot (wasting fuel) or too cool (causing ice crystals at top), it may be possible to remedy the situation by substituting tin stovepipe for insulated sections, or vice versa. Tin stovepipe will radiate heat, while the insulated pipe won't—therefore, the more tin pipe, the more heat will be radiated. A metal chimney can be suspended directly over the stove, eliminating two draft-impeding bends. It also comes in a variety of sizes to match your stove or fireplace.

Insulated metal chimney pipes can be installed closer to combustibles than non-insulated pipe. Check the manufacturer's specifications for exact clearances. Insulated metal chimney pipes come in three varieties: asbestos, fiber-insulated, and triple-wall metal pipe, either with dead air or ventilated.

Prefabricated chimneys are easier to erect than masonry ones. Tests at the National Bureau of Standards have shown that metal and masonry chimneys differ little with respect to draft when used under similar conditions. A key point is that metal prefabricated chimneys must be UL-listed as "All Fuel" chimneys. Do not use the UL-listed "Vent," as it is not insulated or ventilated enough for coal burning. The standard sections are 18 inches and 30 inches long and are available in a variety of inside diameter sizes. The sections lock together and no screws or special tools are needed for assembly.

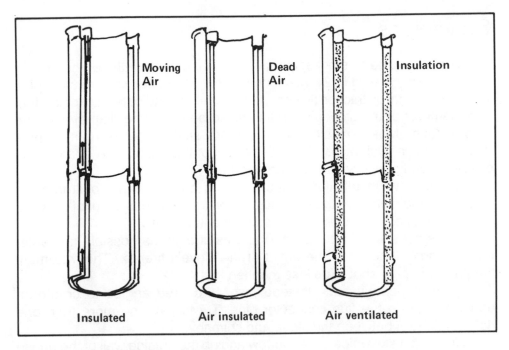

Fig. 48—Types of prefabricated metal chimneys.

Prefabricated metal chimneys should be of the double-wall insulated type, not of the triple-wall, air-cooled type. Insulated pipe retards heat flow outward while keeping coal flue gases hot. Air-cooled pipes cool flue gases, which may reduce draft and increase the rate at which deposits of soot and ash accumulate in the pipe. If an existing chimney is used, it must be inspected to be sure it is in good condition.

The ventilated types can be seriously damaged in a chimney fire. Because the chimney is insulated, temperatures up to 2,200 degrees can occur inside the flue. This can cause buckling and separation of the inner liner, causing the insulation to settle and fall out, thus creating a dangerous hot spot in the chimney. This type of chimney should be kept very clean and inspected thoroughly after a chimney fire.

There is no "best" chimney for all situations. An interior masonry chimney is best for overall efficiency and additional heat storage. Insulated metal chimneys accumulate less creosote and, because they burn hotter, provide a better draft. Triple-wall pipe and lined masonry chimneys are safer during a chimney fire.

Prefabricated insulated chimneys, like all chimneys, should be kept clean to prevent soot, fly ash, and creosote buildup. Tests show that the inner liner can seriously deteriorate because of chimney fires as well as chemical reaction of flue gases. Be sure to check the prefab insulated chimney for any damage if it has endured a chimney fire.

Stovepipes

The stovepipe or smokepipe used to connect the outlet of the firebox to the chimney is sold in 24-inch lengths. Building codes require stovepipe to be 24-gauge or thicker (lower gauge numbers indicate thicker metal). The diameter of the stovepipe used should be the same diameter as the firebox outlet. Most coal stoves use a 6-inch smokepipe. Using stovepipe that is smaller in diameter than the firebox outlet will reduce combustion efficiency and possibly cause improper draft.

Some stove installations should have a damper either built into the stove or in the pipe near the stove to control draft and loss of volatile gases. Check the recommendation of the stove manufacturer.

Stovepipes should be as short and as straight as possible and enter the chimney higher than the outlet of the stove's firebox. The maximum length of the pipe should be less than ten feet.

Avoid horizontal runs. Instead, use 45-degree angles to create an upward slope in the flue connector pipe. Try to have no more than one right angle turn between the stove and chimney.

Running a stovepipe out a window and up the outside wall of the house is a dangerous practice. Most coal stove manufacturers do not recommend long spans of single-thickness stovepipe as a heating device. This idea had some merit when used with inefficient wood stoves where much of the heat went up the pipe. Airtight coal stoves, however, are more efficient and this practice may cause a backdraft situation.

Long stovepipes and those with restrictions should be cleaned frequently to prevent creosote buildup and possibly chimney fires. The entire length of the smokepipe must be easily inspected, firmly fastened at the joints, and kept free of all combustible materials.

Where a smokepipe must pass through an interior wall, provide an opening with at least six inches of clearance from all wood framing and protect it with a double-wall ventilated thimble. This type of thimble is not readily available, but can be fabricated by a sheet metal shop. Ventilation through the thimble is an essential aspect of its design; the ventilating holes on either side must not be blocked.

A thimble about two inches larger than the pipe is used for the installation of a flue for a gas furnace and is not adequate for a wood or coal stove installation.

Stovepipe and chimneys, although they are often mistakenly used interchangeably, are two completely different things. When a stovepipe is used for a chimney dangerous conditions may occur. Ash may build up rapidly, and the wind and rain may soon corrode the pipe. Consequently, if a chimney fire should occur, there will be little to contain it. Chimneys, in contrast, keep the smoke and gases hotter to prevent the buildup of ash and/or creosote, and most chimneys are able to contain a fire.

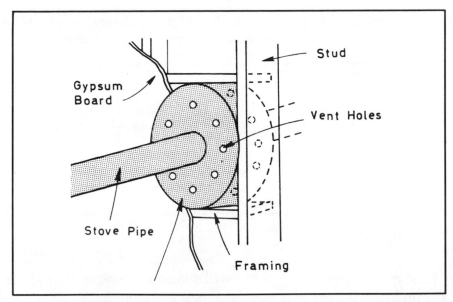

Fig. 49—Ventilated thimble diameter should be three times the diameter of the stove pipe.

A stovepipe must be installed so that there is a good, controlable draft to carry the hot gases away quickly and safely. This can be a tricky proposition unless you follow these guidelines carefully:

1. Keep the stovepipe short.
2. Keep turns and bends to a minimum.
3. The horizontal portion of the stovepipe should be no more than 75 percent of the vertical portion.
4. The stovepipe should enter the chimney well above the stove outlet. Horizontal portions of the stovepipe should rise at least one-quarter-inch per foot.

Stovepipe comes in different diameters and gauges (thicknesses). Measure the pipe connector on the stove and get the proper size stovepipe to match. Pipe gauge 24 or thicker is a good investment because it will last longer (lower gauge numbers indicate thicker metal). The stovepipe pieces should be fitted together tightly, then permanently joined with two or three metal screws to prevent the stovepipe from being shaken apart during a chimney fire. Since stovepipe cannot be expected to last long, inspect it regularly and plan on replacing it every two or three years.

All stoves should have some sort of damper. Dampers are used to control the air flow to a stove, and in the event of a chimney fire, they should have the capability to shut off the airflow to the chimney completely. Generally, airtight stoves have sufficient dampers built in, but any non-airtight stove should have a damper in the stovepipe. It is a good idea to install a second damper in the stovepipe as a safety control. Most of the dampers sold in hardware stores are not solid, and although they can be used to control air flow, they cannot shut it off completely. You may have to replace this damper with a solid piece of metal cut the same size.

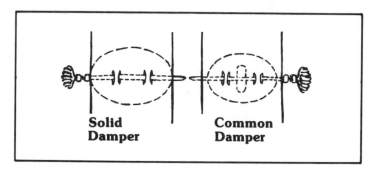

Fig. 50

Passing a stovepipe through a wall should be avoided if possible. If it can't be avoided, special precautions must be taken.

Wall Thimble

Stovepipe should never run through concealed places like closets or attics. Care must also be taken when attaching the stovepipe to a chimney. Stovepipe clearances must be maintained from all combustible surfaces. In a masonry chimney, a special connector thimble must be installed. The stovepipe should fit into this thimble, and the thimble must fit into the chimney and be permanently cemented in place. Do not extend the thimble into the flue. Typical fittings are diagrammed in Fig. 52.

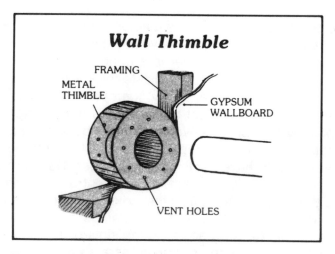

Fig. 51—Wall thimble. Stovepipe should never run through concealed places like closets or attics.

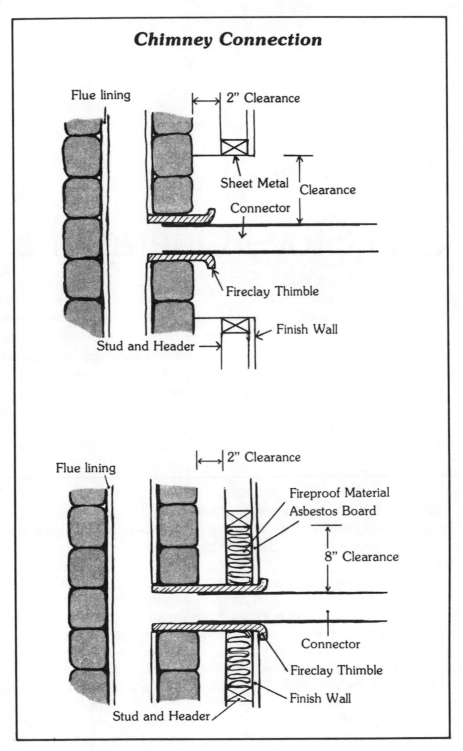

Chimney Connection

Fig. 52—Chimney connection

Chapter 9

Coal Stove Operations

> The suffering man ought really to
> consume his own smoke; there is
> no good in emitting smoke till you
> have made it into fire.
>
> Thomas Carlyle

Burning Coal

Use anthracite or a good quality bituminous coal (soft coal) or brown coal in your stove. Determine the right size coal for the heater, i.e., chestnut, pea, or some other size. The coal must be dry to burn well and should be stored indoors.

Coal is not as easy to burn as wood, and the ease of burning varies with different types and makes of coal stoves, furnaces, and boilers. Burning coal requires patience, and a very specific and regular procedure of loading, shaking, and adjusting. If you do not follow the correct procedure, the coal fire will go out. This can happen in a very short space of time, and once the extinction process has started it is almost impossible to reverse.

After a coal fire goes out, all the coal must be emptied from the stove and the complete starting process must be repeated. The coal-burning learning process is often long and frustrating, but once the proper procedure is established and followed, coal burning becomes a reasonably simple process, with the benefits of long burn times and evenness of output over the entire length of burn.

Start the coal fire by placing crumpled newspaper on top of the grate. About six full-size newspaper sheets should be adequate. Add a layer of kindling wood on top of the paper. Make sure to arrange the wood so that air can pass between the pieces. Wood packed tightly is difficult to ignite. Be sure that the draft damper(s) in the stovepipe is (are) open. After lighting the crumpled newspaper, leave the ash door open for two or three minutes or until the fire has caught well.

As the kindling is burning hot, keep the draft control fully open to establish a hot fire. (The draft control can be of different types and in different locations on each type of stove.) On boilers and furnaces with automatic draft controls, the ash door may be opened for start-up.

Set the ash door damper open fully and close the ash door. If the stove puffs smoke at any time, closing the top damper fully, if it is not already closed, will stop it. After about five minutes, carefully open the top door. The wood inside should be well ignited. If so, allow the wood to burn for fifteen minutes or until it has turned to wood coals. Now, place a thin layer of coal (about one inch to two inches deep) in the stove. Close both doors, open the bottom damper fully and the top damper about halfway.

Continue adding small amounts of coal until there is a solid bed of *burning* coal. Do not add too much at one time. Allow sufficient time between each small loading (at least 10 to 15 minutes) so that each loading has time to thoroughly ignite before the next load is put in. For maximum burning efficiency, *always* fill the stove to the highest level possible. Be sure to follow the directions by the stove manufacturer in order not to overload the heater. A deep bed of coal always will burn more satisfactorily than a shallow bed.

When most of the wood coals are burned and most of the coal is completely ignited, the draft control should be turned down. If the ash door

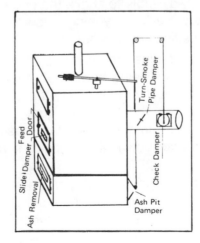

Fig. 53

has been opened, this *must* be closed to prevent overfiring, which can severely damage the stove, boiler, or furnace.

When all the coal is ignited thoroughly, or there is a substantial depth of hot coals, the stove may be shaken thoroughly. After shaking, keep the draft control open until the fire continues to burn hot, then turn the draft control down until the proper level of fire is reached. If the ash door has been opened, be sure to shut it. Serious damage can result if the stove is run for extended periods with the ash door open.

On stoves with screw-type draft controls, count the exact number of turns from full shut to the normal operating positions in order to adjust the stove to the exact level of heat output and effective length of burn.

During the day control the fire with the check damper and ash pit damper. For more heat, open the ash pit damper and loosen the check damper. To reduce the speed of burning, reverse the positions of these dampers. The smokepipe or turn damper should be left as nearly closed as possible. It should not be used in daily regulation of the furnace. The correct position can be determined by experimenting and observing the burning speed of the fire under different settings.

At night, before banking the fire, shake the grates gently until a red glow appears in the ash pit. Pull the "live" coals forward to the bottom of the feed door, leaving a red spot of burning coals exposed. This "red spot" ignites the gases coming from the fresh coals. Add two or three shovels of anthracite coal to last overnight, heaping the coal well to the rear of the fire pot. Check dampers. Do not check the fire to the extent that the house becomes chilled during the night.

In the morning, to bring up the fire, close the check damper and open the ash pit damper. Leave the slide damper in the fire-door slightly open at all times. Add a shovel of coal, leaving a red spot of glowing coal visible near the feed door. It is advisable to shake the grate in the morning even though this was done the previous evening. Allow the coal to burn with the dampers in the open position until the house is entirely heated.

After a few minutes, a blue fire should be coming through the coal. These are the volatile gases being burned off; that is, the coal is being converted to coke. The objective is to ignite the coal and to burn off the gases. Roughly 20 minutes should be adequate, but be careful not to overheat the stove. It is very important to burn these volatiles completely before closing the dampers down. It is important *never* to close either damper fully.

Once the coal is ignited, more coal can be added. The coal in the stove should be burning well before loading more coal. It's a good idea to open the bottom damper fully for five minutes when adding more coal.

Coal never should be added unless there is a reasonably hot fire. If the fire is burning hot and there is a deep bed of coals, full loads of coal can be added at any time. However, if there is not a deep bed of coals, it is best to add small amounts of coal at first.

The coal stove is equipped with movable grates called shaker grates. A bed of burning coal normally should not be disturbed from above as this causes clinkers (fused ash and unburned coal) to form. You can even lose the fire by disturbing it too much from above. In the morning and in the evening, preferably just before loading, shake the grates. Use a short, choppy motion and only shake them enough to see a red glow begin to appear above the grates when viewed through the ash door, or until a few red coals fall. The object is to leave an insulating layer of ash on top of the grates to protect them, yet remove enough ash to allow proper air flow through the coal.

After the fire is started and thoroughly ignited, for average daily operation, control the fire with the check damper and ash pit damper. If a moderate fire is required open the check damper wide and close the ash pit damper. As additional heat is required, close the check damper partially or fully and open the ash pit damper partially or fully.

Before coaling the fire for the night, shake the grates gently until a red glow appears in the ash. Do not shake hard (by shaking moderately a bed of ashes remains on the grates). Wet ashes thoroughly to eliminate dust and remove just before shaking the following night. To check the fire, close the ash pit damper completely and open the check damper to suit weather conditions.

With chestnut, pea, and buckwheat be sure to leave a red spot of glowing coal visible. This will ignite the gas created from the fresh charge of coal and insure maximum heat from the fuel. In the morning, increase draft by closing the check damper and opening the ash damper. When the fire is burning brightly, add coal.

Burning coal is very easy after a few days of practice. Coal produces more ash than wood. This is because coal is about eight percent inert matter that will not burn, as opposed to about one and one half percent for wood. Any coal or very hard pieces may be sifted out of the ash and used again. Never force the shaker handle. Should the grates ever get jammed while the stove is running, either let it run for a while to burn away the hard chunk jamming it, or use a hooked poker from below. Ashes should be removed before shaking the grates in order to avoid removing hot coals. Always keep the ash can tightly covered and store it on a non-combustible surface.

Shaking the coal grates should be done only with a hot stove, at least once but not more than twice a day. Best results from shaking will occur if short, "choppy" strokes are used rather than long, even strokes. The amount of shaking is critical. Too little or too much can extinguish a fire due to blocked air flow. The proper amount normally occurs when red coals first start to drop through onto the bed of ashes.

Coal ashes should never be allowed to accumulate in the ash pit so that they in any way impede the flow of combustion air to the fire. Excess ash accumulation can cause the fire to go out, and can also cause severe damage to the grates because they cannot cool from a flow of air beneath them.

Clinkers can occur in any coal stove. They are pieces of fused ash that are hard. They can become large, and therefore cannot be shaken through the grates in a coal stove. When there is an appreciable accumulation, the fire will go out because insufficient air is allowed to pass through the clinkers to the burning coal. Once clinkers have formed, they can be removed only from above the grates. This usually means the fire must be allowed to die out before they can be removed.

Clinker formation can occur from a number of different causes or a combination of causes. Some of these are as follows: too hot a fire (too much draft); too shallow a bed of coals; too deep a bed of coals; excess shaking, particularly with rocker grates; poking the fire from the top; poor quality coal (excess ash content); too little air (draft) after a long hot fire.

Coal Stove Safety

Whenever a loading door or lid is opened, it *always* should be cracked slightly before fully opening to allow oxygen to enter and burn any combustible gases that are present. Failure to do this could result in sudden ignition of the unburned gases when the door is opened.

A stove, furnace, or boiler never should be filled with excess coal so that the flue gas exit is in any way blocked or impeded. Burning coal generates carbon monoxide. If the flue gas exit is blocked, the carbon monoxide can be forced out of the stove into the room, with possible fatal consequences.

With the exception of the start-up period, an ash pit door never should be left open. Serious damage to the stove can occur from overheating and, in extreme cases, this overheating could be a possible cause of an "unfriendly fire."

Coal gas can seep into the living space through cracks in a stove or flue pipe. Be sure your coal stove is in good repair, and *never* burn coal in any stove that does not have an airtight, unified chimney system.

Coal-burning appliances should be used only with chimney systems that provide a strong, reliable positive draft. Users are advised to prevent back-puffing of coal smoke into the living space. **Remember: Coal gases are toxic.**

Some Do's and Don'ts of Coal Burning

DO   Remove ashes daily from under the grate. If they are allowed to build up, damage could occur to the grate.

DO   Dispose of ashes in a metal container. Always cover the container and place it well away from any combustible material.

DON'T Overfire the heater. Overfiring is evident when any part of the stove becomes red. Besides voiding your warranty, this practice could be dangerous.

DON'T Add wood or paper to the coal bed. Ash may fuse into what is known as a clinker and put out your fire.

DON'T Stir or rearrange a coal fire, as this may cause a clinker problem.

DON'T Use lighter fluid, kerosene, gasoline, or any such flammable liquids to start a coal fire.

DON'T Bank coal against a door, as spillage of hot coals may occur.

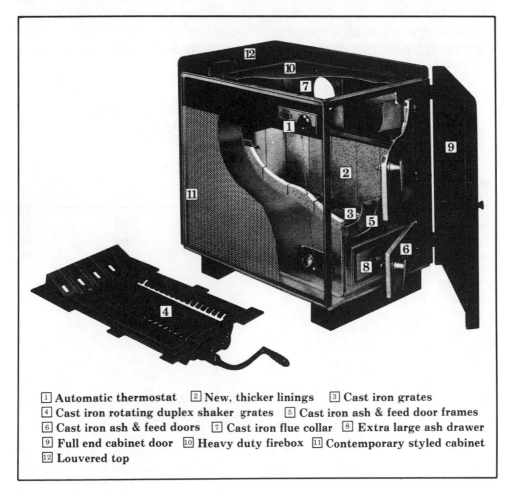

1 Automatic thermostat   2 New, thicker linings   3 Cast iron grates
4 Cast iron rotating duplex shaker grates   5 Cast iron ash & feed door frames
6 Cast iron ash & feed doors   7 Cast iron flue collar   8 Extra large ash drawer
9 Full end cabinet door   10 Heavy duty firebox   11 Contemporary styled cabinet
12 Louvered top

Fig. 54—King Coal circulator

Ash

Refuse from the stove or furnace consists of ash, coked carbon, and unchanged combustible matter from the coal. The weight of refuse actually formed per unit weight of coal may be calculated, since all the ash content in the coal, as reported in the proximate analysis, goes into the ash pit as part of the refuse.

Combustion of the mineral matter in coal produces a residue called ash, and ultimate coal analysis shows a corrected ash content. The composition of ash is important to the coal stove owner because of the relationship of the chemically component parts (i.e., ultimate analysis of coal) to each other in explaining the cause of clinkering.

After all the combustible matter in the coal has been burned off, the residual mineral matter can be analyzed to show its component parts by the same chemical methods as those used in rock analysis. All of the elements found in coal ash are fully oxidized and are usually reported on a percentage basis as the oxides of the constituent elements. The main constituents with their chemical symbols usually reported are as follows:

TABLE 9
Sample—CHEMICAL ANALYSES AND SOFTENING TEMPERATURES OF COAL ASHES
(Range of percentages)

Constituent	From	To
Silica (SiO_2)	55.00	56.62
Alumina (Al_2O_3)	31.1	38.12
Ferric Oxide (Fe_2O_3)	2.06	10.1
Titanium Oxide (TiO_2)	0.96	1.82
Cupric Oxide (CuO)	0.04	0.09
Manganous Oxide (MnO)	0.03	0.12
Calcium Oxide (CaO)	0.3	1.40
Magnesium Oxide (MgO)	0.0	0.76
Sodium Oxide/ Potassium Oxide (Na_2O/K_2O)	0.44	1.3
Phosphorous Pentoxide (P_2O_5)	0	0.14
Sulphur Trioxide (SO_3)	0.0	0.80
Softening Temp., Degrees F.	2925	3000

Silica, Alumina, and Ferric Oxide are the principal constituents of the ash; the other items are usually found in small amounts but can be important in special applications.

All coal seams contain inert mineral matter that does not burn. This residue left after combustion is called ash. Ash is considered to be inherent or extraneous, depending upon the ease of removing the mineral impurities by mechanical means.

Inherent ash is ash resulting from finely divided mineral impurities that are impractical to remove by mechanical means. Some of the mineral matter is derived from the original plants from which coal was formed. Other finely divided impurities originated in the form of sediment carried into the coal-forming swamps.

Extraneous ash is ash resulting from mineral impurities that can be removed by mechanical means. This includes clays and shales from the floor or roof of the coal seam. Also included are partings (layers of impurities within the seam) and nodules of impurities not finely divided and distributed within the coal structure.

The volume of ash that can be expected from the burning of one ton of anthracite coal ranges from 9 to 12 cubic feet. The ash will probably weigh 25 to 45 pounds per cubic foot. These estimates will vary depending on the percentage of ash produced as a product of combustion. The efficiency of the combustion process, affected by either the stove design or the firing procedures, will determine the amount of carbon or unburned coal that appears in the ash. The greater the amount of carbon in the ash, the greater the weight and volume of the ash. In other words, less inefficient stoves or stove operators will produce more and heavier ash. Adding to this relationship, clinkered ash will be heavier than non-clinkered ash and clinkered ash will take up more room than the fine powdered ash that results from the efficient burning of good quality coal.

Fig. 55—Luneberg Boilerplate Gravity Furnace

Practically speaking, ashes should be placed in a metal container with a tight fitting lid. The closed container of ashes should be placed on a non-combustible floor or on the ground, well away from combustible materials, pending final disposal. If the coal ashes are disposed of by burial in soil or otherwise locally dispersed, they should be retained in the closed container until all cinders have thoroughly cooled.

Smoke

The burning of most solid fuels under certain conditions may produce byproducts containing enough carbonaceous material to make the exhaust gases visible. This visible gas is smoke. With proper or perfect combustion conditions, all solid fuels can be burned without smoke. Most airtight coal-burning stoves operate efficiently enough to produce not very much smoke. Anthracite coal is itself a relatively clean-burning fuel if sufficient air is supplied for combustion and an efficient draft system allows for complete secondary combustion. These conditions are usually present during the operational phase of a coal-burning stove's day. Conditions sometimes deteriorate after a fresh load or change of coal is placed in the stove. Generally what occurs is that the new coal takes some time to reach its ignition temperature and incomplete combustion occurs. Also, the new coal disrupts the previously established air patterns in the fuel bed and holds the heat down inside the fuel mass. The result is the incandescent coals that are capable of igniting the secondary combustion that normally occurs above the fuel bed.

Proper loading procedures can minimize these conditions and significantly reduce or eliminate the production of smoke. A smokeless fire is generally indicative of efficient consumption of coal.

Smoke may also be caused by improper stove or furnace design. Design faults may be corrected by alterations to improve combustion conditions to allow for more temperature and/or turbulence. Before making any substantial modifications to a heating device it is advisable to contact the manufacturer to discuss your proposed changes. Since most manufacturers are conscientious, and most stoves that present serious smoke problems would not be popular sellers, it is more likely that the problem is not in the design but in either the operation of the stove or the draft. The manufacturer can help to trouble-shoot these areas.

More likely than a design problem with a stove is a maintenance problem. Carefully check for leaks in the stove. Improperly sealed seams, doors, or other openings can allow air to disrupt the combustion process or affect the draft.

Draft can be the cause of smoke by creating combustion rates that cause such a large volume of volatile gases that complete secondary combustion is not possible given the interior dimensions of the stove or furnace. Excessive load demand placed on the stove can also create smoke. Insufficient draft caused by clinkers, chimney problems, or leaking stoves may also be the culprits. Last, but not least, the stove operator may be at fault by attempting to operate the stove or furnace at temperatures that are too low for the device's design or by improper loading procedures. Proper operation of a coal-burning stove or furnace, whether hand-fired or stoker-fired, calls for the maintenance of uniform fuel beds. Allowing a stove to nearly deplete its fuel supply before reloading is a common cause of both smoke and inefficient operation.

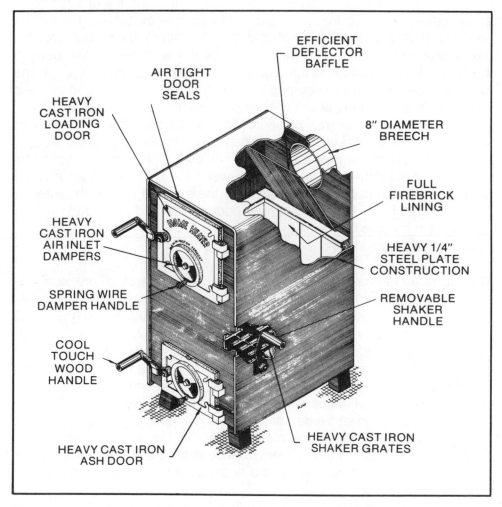

Fig. 56—Home Heater coal stove—Model H1 burns anthracite or bituminous coal (nut or pea size), holds up to 100 lbs., and has 75,000 Btu-per-hour output.

Draft Improvement[1]

An inadequate draft is almost invariably the cause of faulty operation. The signs are that the fuel burns badly or goes out and the heater smokes. In all cases check the chimney, the draft proofing in the stove, and the cleaning access doors. Excessive draft causes the stove to burn too fast and clinkers form when burning coal. Check the air tightness of the primary air inlet and the chimney draw. If the latter is found to be in excess of the recommended optimum depression, it may be necessary to fit a draft regulator.

Draft is the difference in pressure available for producing a flow of gases. If the gases within a stack are heated, each cubic foot will expand, and the weight of the expanded gas per cubic foot will be less than that of a cubic foot of cold air outside the chimney. Therefore, the unit pressure at the stack base due to the weight of the column of heated gas will be less than that due to a column of cold air. This difference in pressure, like the difference in a head of water, will cause a flow of air into the base of the stack. In its passage to the stack, the cold air must pass through the furnace or stove where it becomes heated. This newly heated gas will also rise in the stack and the action will be continuous.

The intensity of the draft, or difference in pressure, is usually measured in inches of water. Assuming an atmospheric temperature of 62° F. and the temperature of the gases in the chimney as 500° F. and, neglecting for the moment the difference in density between the chimney gases and the air, the difference between the weights of the external air and the internal flue gases per cubic foot is 0.0347 per pound, obtained as follows:

Weight of a cubic foot of air at   62° F. = 0.0761 lb.
Weight of a cubic foot of air at 500° F. = 0.0414 lb.
                          Difference = 0.0347 lb.

Therefore, a chimney 35 feet high would have a theoretical pressure exerted on each square foot of its cross-sectional area at its base of 0.0347 × 35 = 1.215 pounds. As a cubic foot of water at 62° F. weighs 62.32 pounds, an inch of water would exert a pressure of 62.32 ÷ 12 = 5.193 pounds/square foot. The 35-foot stack would, therefore, under the above temperature conditions, show a draft of 1.215 ÷ 5.193 or approximately 0.234 inches of water. Actually, restrictions in the chimney and friction, as well as the cooling of gases, will reduce the figure materially.

The rapid union of oxygen and carbon is known as combustion. The result of this reaction is the liberation of heat and the production of carbon dioxide. It is necessary that sufficient draft be available to supply the oxygen needed to support combustion at the desired rate. The draft is measured

[1]The Anthracite Institute Laboratory Notes, Chapter IV, "Draft Difficulties" is acknowledged.

with a draft gauge of the type furnace servicemen use, and is measured with the stove burning and the chimney warm. Take the measurement between the flue pipe fitting on the stove and the barometric damper (if used) in the flue pipe. A barometric damper will reduce the draft if it is too strong. It will also prevent wind from increasing the draft of your stove, preventing unintentional overfiring and possible damage. Barometric dampers must be installed with the opening facing away from walls.

The automatic draft control simplifies tending to the coal fire and saves fuel. There is a heat sensor by the flue pipe fitting which monitors the temperature of the flue gases and responds quickly to adjust the combustion air. This is calibrated with numbers instead of temperatures because fuel type and many other factors determine what setting will give what temperature. Rough handling may damage this control. Otherwise, this unit should be trouble-free for many years.

The theoretical importance of excess air and draft for efficient coal-burning operation has already been discussed. In order to burn any solid fuel there must be enough draft to insure proper directional air flow and sufficient air for proper combustion. Draft is the upward movement of air through a chimney flue. It is a function of the temperature difference between the warm inside house air and the cooler outside air temperature.

Experience shows that the majority of heating difficulties are traceable to draft. There is little excuse for this and it would not be so if proper attention was always given to the location, construction, and maintenance of chimneys. The causes of poor draft and the remedies for it are many.

When the cross-sectional area of an existing or new chimney is too small, it will be impossible to get quick pick-up or rapid burning, and the result is insufficient heat. If the stack is too large in a cross-sectional area, gases expand and cool, causing a reduction in draft. This may be corrected by installing an induced draft fan if changes to the stack are too costly.

A chimney must always clear the top of the building it serves as well as adjacent buildings or trees; otherwise a downdraft condition will be experienced. A dangerous downdraft is the reversal of normal air direction and can cause a carbon monoxide hazard.

Major Causes for Poor Draft

1. TOP OF CHIMNEY TOO LOW—SHOULD EXTEND ABOVE TOP OF ROOF AND SURROUNDING TREES
2. CAP ON TOP OBSTRUCTS DRAFT OR ENCOURAGES NESTING BIRDS
3. CRACKS IN THE BRICKWORK
4. CHIMNEY HEIGHT INSUFFICIENT TO DEVELOP NECESSARY DRAFT

 5. CHIMNEY MISSING BRICK IN DIVISION WALL
 6. JOIST PROTRUDING IN CHIMNEY—RESTRICTS AREA AND CONSTITUTES SERIOUS FIRE HAZARD
 7. BRICK WEDGED IN THE CHIMNEY
 8. CHIMNEY NEEDS CLEANING
 9. OPEN JOINTS ALLOW LEAKAGE
10. POOR CHIMNEY DESIGN AND CONSTRUCTION OF OFFSET CAUSE DRAFT LOSS
11. DEPOSIT OF DIRT AND SOOT IN CHIMNEY
12. OBSTRUCTION OF THE FLUES
13. THROAT OF FIREPLACE TOO NARROW—CAUSES SMOKING
14. AIR LEAKS AROUND PIPE OR PIPE PROJECTS INTO CHIMNEY
15. FLUE OUTLET NOT CLOSED
16. FURNACE PIPE TOO LONG—FILLED WITH SOOT AND SLANTS DOWN
17. CLEANOUT DOOR LOOSE

The chimney cleanout must be tight or cold air will be drawn in between the cleanout door and frames and will cool the stack gases and reduce the available draft. The cleanout door should be tight-fitting in order to avoid loss of draft. Leaks between bricks must be stopped. Use a flame test to find air leaks. The flame will be pulled into the smokepipe wherever a leak is present. Draft leaks are easily detected around the furnace or smokepipe by using a lit candle or piece of paper and holding it near any point where a leak is possible. If the flame is deflected, a leak is present.

The stove smokepipe should be of a cross-sectional area equal to that advocated by the stove manufacturer and as nearly round or square as possible. These shapes allow for a minimum of frictional resistance which cuts down effective draft.

A smokepipe must be tight at all joints and free from holes, cracks, or other openings. The check damper should fit the smokepipe closely and tightly. Use the flame test with candle. Cracks and small holes may be filled with asbestos or iron cement. If it is beyond repair in this way, replacement of the pipe should be made. Smokepipes must be free from soot or other obstructions which decrease the cross-sectional area of the pipe and, therefore, cut down draft. Stove and furnace fronts, sections, and doors must fit tightly. Chimneys in general must be tight; otherwise cold air will be drawn in between loosely set bricks and will cool the gases in the chimney, thus reducing draft. To test the tightness of a chimney, block off the top and light a smoky fire of wet wood, waste, or rags in the boiler. If smoke escapes through the chimney to the inside of the house or outdoors, the stack should be repaired and made tight.

Some Requirements for Good Draft

1. CHIMNEY TOP HIGH ENOUGH—SHOULD EXTEND ABOVE TOP OF ROOF AND SURROUNDING TREES AND PROVIDE FREE DISCHARGE OF FLUE GAS
2. MORTAR JOINTS POINTED PROPERLY
3. TILE LINING IN ALL FLUES
4. DIVISION WALL COMPLETE
5. JOISTS FREE FROM CHIMNEY—AND CHIMNEY IS NOT USED TO SUPPORT BUILDING
6. FLUES CLEAR AND CLEAN
7. JOINTS OF FLUE LINING SMOOTH AND TIGHT WITH ADJACENT FLUE JOINTS STAGGERED
8. CHIMNEY OFFSET GRADUAL
9. FIREPLACE THROAT FULL-WIDTH—PREVENTS SMOKING
10. PIPE CONNECTION FLUSH WITH FLUE-JOINT SEALED
11. ALL FLUE OUTLETS CLOSED
12. SHORT FURNACE PIPE USED AND UPWARD SLANT DAMPERS IN PROPER ORDER
13. CLEANOUT DOOR TIGHT

The inside of a chimney must be free from obstructions; generally from mortar, brick, and tin, which partially restrict the area of the chimney and thus retard flow of gases up the stack, reducing the draft.

A chimney should be of uniform cross-sectional area throughout its length. A chimney should be straight and free from offsets and bonds. Offsets, if necessary, should be made with a gentle curve.

A smokepipe must not project into the inside of a chimney flue; this can coke the flow of gases from the stove and may severely restrict the draft. Smokepipes should be straight and as short as possible, with an upward pitch.

Unless unavoidable, a chimney should serve only one heater. Do not connect a coal stove to the same flue serving a fireplace, because sparks and flue gases from the stove may enter the house through the open fireplace. Where two units are connected to one chimney, means should be provided for effectively sealing either one or the other against draft leaks when one is not in use. This can be done with a sliding damper such as is used for a blast gate or a blower. If, despite these recommendations, two stoves are connected to the same chimney, the connections must enter the chimney at different elevations.

If the chimney is serving one of the many wood/coal combination stoves or furnaces, it should also be cleaned of any creosote that has accumulated as a result of the wood-burning portion of the heating season. It will be best to clean the chimney in the spring.

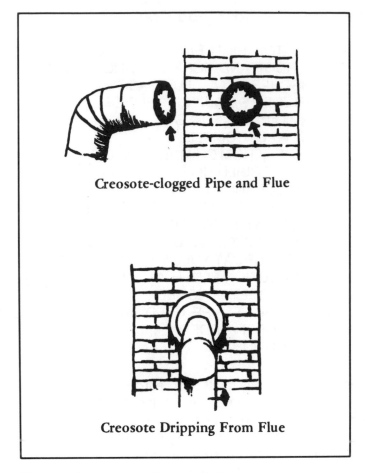

Creosote-clogged Pipe and Flue

Creosote Dripping From Flue

Fig. 57—Creosote problems

Chimney Fires

If you should build a fire which keeps getting hotter and hotter, or if the coal stove starts to glow pink or red, you have either too hot a fire or a "runaway" fire.

Chimney fires for coal heaters are possible in all but the cleanest chimneys. Flames may leap high above the chimney top, carrying burning particles that can set fire to the roof or any surrounding combustibles. Fire may also escape through flaws in the chimney direction to the structural house. Firemen know best how to protect your home and prevent any possible spread of the fire. It is wise to have safe portable ladders nearby for access to roofs if organized fire protection is remote or unavailable.

If a chimney fire occurs, follow these precatutions:
1. Call the fire department.
2. Close all air inlets and dampers to try to smother the fire.
3. Use a fire extinguisher pointed into the stove. CO_2 extinguishers can be used, or use a chimney fire extinguisher, a flare-like stick that is ignited and thrown into the stove.
4. Wet down the roof and adjacent areas to prevent the fire from spreading.

Chimney fires occur when large deposits of creosote are exposed to very high temperatures. The fire begins with a crackling sound, and as the fire intensifies the stovepipe may shake, turn red hot, and roar like a blow-torch. A plume of fire may rise out of the chimney, and sparks and burning debris may spread the fire to adjacent roofs and walls. The best way to control a chimney fire is to cut off the air going to it. If you have made sure your chimney is sound and airtight, and if the stovepipe and stove connection are tight and secure and if there is only one stove connected to the flue, you can control the air by shutting the air inlet or the solid flue damper. The importance of proper stove installation is obvious.

Don't assume that once the fire has burned out your problems are over. After a chimney fire, a chimney should be checked carefully before it is used again. The high temperatures can crack the liner and violent forces can shake it apart. Without careful inspection, a second chimney fire may finish the job the first one started.

Chimney cleaning can be very hazardous to your health. People have fallen off roofs. Recently, coal tar residues have been suspected of causing some types of skin cancer. Avoid breathing fly ash dust and scrub thoroughly after finishing the chimney cleaning.

Proper equipment for these cleaning assignments may be obtained from quality chimney sweep suppliers. If the task sounds too formidable, the cleaning may be contracted to a reputable chimney sweep. In any case, do not ignore the chimney, for it has a habit of presenting problems at the most inconvenient times, from small nuisances to life-threatening situations.

Chapter 10

Coal Fuel

No need of "Carrying Coals to New Castle"
James Melville, 1583

We have the largest recoverable coal reserves in the world—about one-third of the total. Approximately one-eighth of the area of the United States is underlain by coal-bearing strata, and these strata occur in at least 37 states. Coal is available now and more than sufficiently plentiful to meet burgeoning energy demands. In light of the growing energy shortage, coal is the key to our ability to get through this century and into the next.

The best type of coal to burn in coal heaters is anthracite. It is a hard coal with a shiny surface and contains the highest percentage of carbon and the lowest amount of volatile matter of the many types of coal. It does not contain much sulfur and produces very little smoke. It burns with a short blue flame and an intense, steady heat.

There are three popular sizes of coal that are generally recommended for use in coal hand-fired heaters. They are pea coal (9/16″ to 1-3/16″), chestnut coal (1-3/16″ to 1-5/8″), and the largest size, stove coal (1-5/8″ to 2-7/16″). The most recommended size is the middle size: chestnut coal. You should closely follow the recommendations of the manufacturer of the stove you purchase when it comes to the size of the coal you use. Most manufacturers have tested their stoves extensively and you can gain much by following their instructions.

While local conditions and manufacturers' suggestions will influence the proper size of fuel to be used, the following guidelines may be helpful.

Egg is the largest size of domestic Anthracite and should be used in fire pots having a diameter or width of not less than 24 inches, and a depth of at least 16 inches. The firing of this size in smaller stoves, while justifiable in some instances, often results in an unncessary ash-pit loss.

Stove coal is generally suitable for domestic heating plants where the fire pot is not less than 16 inches wide and 12 inches deep.

Chestnut is suitable for any stove or furnace having a firebox 10 to 16 inches deep and up to 20 inches in diameter. Chestnut is also ideal for many types of kitchen ranges and service water heaters.

Pea coal frequently can be used to advantage when the boiler or furnace is considerably larger than necessary. This size can also be used in mild weather and for banking. Pea size can sometimes be substituted for the larger sizes without change to the stove, providing extra care is used in shaking the grates, and adequate draft is available. It is also an excellent fuel for service water heaters and kitchen ranges. Many self-feeding European stoves require this size as larger pieces will not pass through the loading mechanism.

No. 1 buckwheat is the smallest size that can be burned with natural draft. It is not recommended for use where the chimney is less than 50 feet high, or where the stove may be overloaded. Although it is possible to fire buckwheat on large mesh grates by carrying a layer of ashes below the active fuel bed, it is preferable to install fine mesh grates. It is frequently used in magazine feed boilers, mechanical burners, and with forced draft blowers.

No. 2 Buckwheat or Rice Anthracite in domestic heating is used only with mechanical stoking devices.

No. 3 Buckwheat or Barley has no application in domestic heating, but is used extensively in manufacturing plants in connection with chain grate stokers, where it is both economical and absolutely smokeless.[1]

TABLE 10
THE PROPER SIZE OF COAL

Size of Coal	Test Mesh-Round Through	Over	Oversize Max.	Undersize Max.	Min.	Slate	Bone		Ash
Broken	10"	3 1/4"		15%	7 1/2%	1 1/2%	2%	OR	11%
Egg	3 1/4-3"	2 7/16"	5%	15%	7 1/2%	1 1/2%	2%	OR	11%
Stove	2 7/16"	1 5/8"	7 1/2%	15%	7 1/2%	2%	3%	OR	11%
Nut	1 5/8"	13/16"	7 1/2%	15%	7 1/2%	3%	4%	OR	11%
Pea	1 3/16"	9/16"	10%	15%	7 1/2%	4%	5%	OR	12%
Buckwheat	9/16"	5/16"	10%	15%	7 1/2%				13%
Rice	5/16"	3/16"	10%	17%	7 1/2%				13%
Barley	3/16"	3/32"	10%	20%	10%				15%
No. 4 Buck	3/32"	3/64"	20%	30%	10%				15%
No. 5 Buck	3/64"		30%	No Limit					16%

[1]From the Anthracite Institute.

Properties of Coal

Anthracite or hard coal has a slightly lower heating value than bituminous or soft coal. According to engineering data, anthracite has a heating value of 25,400,000 Btu's, while bituminous is 26,200,000 Btu's per short ton (2,000 pounds). When the heating value of anthracite coal is compared with other types of fuel, the equations are as follows:

2,000 pounds of coal = 183.96 gallons of oil[2]

2,000 pounds of coal = 1 to 1.2 cords of seasonal hardwood

Anthracite (hard) coal, which burns with little smoke, comes in many sizes. The type of unit the coal will be burned in determines the size that should be used. For a list of these sizes, see Table 10.

Bituminous (soft) coal is predominantly used by industry. It produces large amounts of smoke. This type of coal also comes in different sizes as required to meet any particular industrial need.

Heat value of the different sizes of coal varies little, but certain sizes are better suited for burning in firepots of given sizes and depths. Both anthracite and bituminous coal are used in stoker firing. It should be noted that cannel coal is a variety of bituminous coal of spores, pollen, and other exceedingly minute plant particles, and has a dull luster and a very homogeneous appearance because of the small size of the component particles. Usually cannel coal burns in hand-fired heaters and care must be used to be sure not to overfire (overheat) the stove if cannel coal is used.

Because cannel coal explodes, it probably should only be used in small quantities in a fireplace with a screen. Avoid using it in a coal stove.

Sale of Anthracite Coal

The wholesale price of anthracite coal at the time this book was published was about $50 per ton, F.O.B. at the mine. This is for the egg, stove, and chestnut sizes. Pea, buckwheat, and rice sizes sell for about $5.25 a ton more. Coal bought bagged at the mine is an additional $37 a ton. The coal from Stac Industries is $3.75 to $4.75 per 50-pound bag delivered.

---

[2]Comparison of the heating value of anthracite coal to oil:

2,000 lbs. of bituminous coal = 4.52 barrels of oil

$$\frac{4.52 \text{ bbls}}{26,200,000 \text{ BTU's}} = \frac{X}{25,400,000 \text{ BTU's}} \qquad X = 4.38 \text{ barrels of oil}$$

1 barrel = 42 gallons

4.38 × 42 gallons = 183.96 gallons of oil

Most coal is brought in by rail. Dealers that buy lesser amounts use trucking as transportation. A truck carries 22 to 29 tons of coal, whereas a railroad car will carry 70 to 100 tons. Estimated transportation costs for coal are approximately $18 a ton by rail and $22 by truck. Coal is sold by the ton, loose or bagged, and in 25-, 40-, 50-, 60-, 70-, and 100-pound bags.

Dealers in the state charge anywhere from $90/ton (picked up) to $125/ton delivered. Typical prices for coal in 50-pound bags range from $2.75 to $5.95. The next bag size most commonly sold is the 100-pound bag. Prices for this size average $6; the range is $4.50 to $7.50 per bag.

The most common sizes of coal sold are chestnut, stove, and pea. The most popular by far is chestnut or "nut" size since it is the most recommended for home stoves.

Price Comparisons with Other Fuels

For a consumer trying to determine which fuel, i.e. coal, oil or wood, is the best buy per Btu value, prices at this time show that wood has the advantage.[3] Coal users, as shown by the following equations, experience a 75% savings over oil, but a 33% increase in cost over wood use.

Coal $\dfrac{\$101.91/\text{ton}[4]}{25{,}400{,}000 \text{ BTU's/ton}}$ $= \$.000004/\text{Btu or } 4 \times 10^{-6}$

Oil $\dfrac{96.4\text{¢/gallon}[5]}{138{,}000 \text{ BTU's/gallon}}$ $= \$.000007/\text{Btu or } 7 \times 10^{-6}$

Wood $\dfrac{\$85/\text{cord}[6]}{25{,}000{,}000 \text{ BTU's/cord}}$ $= \$.000003/\text{Btu or } 3 \times 10^{-6}$

Comparison of coal with oil:
$$\frac{(7 \times 10^{-6}) - (4 \times 10^{-6})}{4 \times 10^{-6}} = \frac{3 \times 10^{-6}}{4 \times 10^{-6}} = .75 \text{ or } 75\% \text{ savings}$$

Comparison of coal with wood:
$$\frac{(4 \times 10^{-6}) - (3 \times 10^{-6})}{3 \times 10^{-6}} = \frac{1 \times 10^{-6}}{3 \times 10^{-6}} = .33 \text{ or } 33\% \text{ increase in cost}$$

---

[3]The heating value of any fuel, due to its composition, may vary within wide limits. This is also the case with the efficiency of the stove, another variable not discussed.
[4]Survey of coal dealers selling a loose ton (1980 selling price).
[5]Survey of oil dealers (1980 selling price).
[6]Price of stove length, seasoned hardwood in 1980.

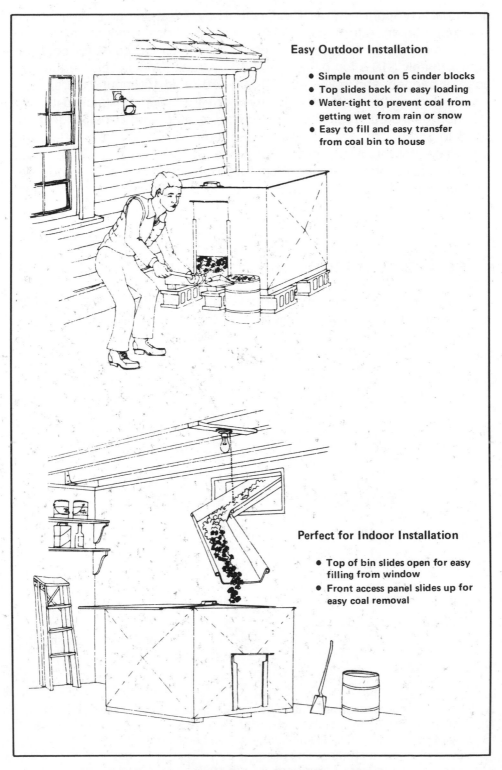

Easy Outdoor Installation

- Simple mount on 5 cinder blocks
- Top slides back for easy loading
- Water-tight to prevent coal from getting wet from rain or snow
- Easy to fill and easy transfer from coal bin to house

Perfect for Indoor Installation

- Top of bin slides open for easy filling from window
- Front access panel slides up for easy coal removal

Fig. 58—Coal bins from Quality Metal Craft, Wollaston, MA.

Coal compares favorably with gas, oil, and electricity in constant heat output because it burns at extremely constant temperatures, while wood stoves may exhibit high temperatures shortly after loading and gradually decreasing temperatures as the fuel is consumed. The use of coal or wood is more work, and requires more attention than gas, oil, or electricity.

The storage of coal is somewhat easier than wood from both a space and handling point of view. A cord of wood, generally, occupies an area 4 feet × 4 feet × 8 feet or 128 cubic feet. A ton of anthracite coal may be sorted in a box or bin 4 feet × 4 feet × 2 feet or 32 cubic feet. Coal may be stored indoors without worry of unwanted insect pests accompanying the fuel.

Bagged coal, while more expensive, may be moved directly to the stove without shoveling and handling, and since most anthracite is washed it is relatively free from dust. It therefore can be stored in a finished basement, family room, or playroom.

In some parts of the country coal is delivered to the home in convenient 55-gallon drums. Many of our older homes still contain coal bins and basement loading chutes. Bins can be constructed of plywood and framing lumber or one-inch tongue-and-groove lumber.

Coal bins can be made to easily fit in a two-car garage, built in the cellar, or stored outside. For less than $60, using 4 feet × 8 feet standard building materials and two hours labor, a three-ton capacity coal bin can be made. Coal bins are available in sheet metal; these commercial units hold less coal than a wooden bin and are four times more expensive than the home-made wooden variety.

Coal can be burned directly as it is received from the supplier and needs no seasoning or drying. Moisture generally does not affect it but it should be protected from rain and snow if it is stored outdoors. Rain and snow can freeze on and among the pieces of coal, making removal from the pile difficult at best, and any moisture clinging to the pieces will have to be burned off in the stove before the fuel is consumed, thereby reducing the operating efficiency of the stove.

Chapter 11

Domestic Water Heating and Automatic Coal Boilers

Now fades the glossy, cherished anthracite;
The radiators lose their temperature:
Now ill avail, on such a frosty night,
The short and simple flannels of the poor.
Christopher Morley, *Elegy
Written in a Country Coal-Bin*

Hot Water System

The average family spends about $275 for hot-water heating, which should provide some information about the payback time on any conversion. After three years the coal water-heating system will be fully paid for, the only cost being coal for the heater. Different types of water-heating equipment and automatic coal boilers are available for heating the home.

Considerations in selecting a coal unit include heating requirements, installation and maintenance costs, and heating costs. Coal heating-equipment dealers and contractors can be of assistance in determining hot-water-heating requirements and in selecting the most efficient, economical (manual or automatic) coal boiler for the home heating system.

For safety and efficiency, a reputable coal heating contractor should be engaged to install the central heating system and inspect it once a year, including cleaning of the unit and chimney. A less costly hot water system correctly installed will be more satisfactory than an expensive one that is not the right size for the house or that is not properly installed.

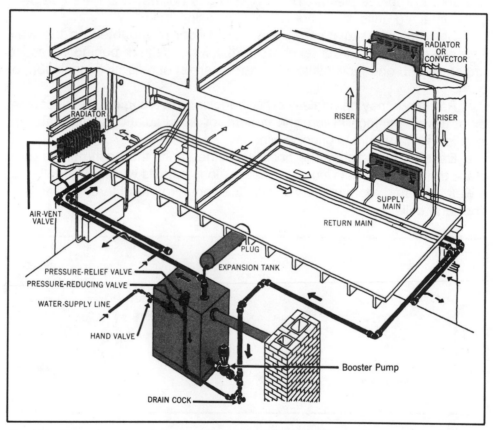

Fig. 59—Two-pipe forced-hot-water systems have two supply pipes or mains. One supplies the hot water to the room heating units, and the other returns the cooled water to the boiler.

Area heating units, including radiant stoves, circulator heaters, and "pipeless" furnaces, are installed in the room or area to be directly heated. In central (domestic hot water) systems, the heating unit is usually located in the basement or other out-of-the-way place and heat is distributed through pipes. Although stoves (coal and wood) are cheaper than central heating systems, they are dirtier, require more attention, and heat less uniformly. Central heating systems (hot water and others) are the most efficient and economical method of heating.

A domestic hot water and steam heating system consists of a boiler, pipes, and room heating units (radiators or convectors). Hot water or steam, heated or generated in the boiler, is circulated through the pipes to the radiators or convectors where the heat is transferred to the room air. Boilers are made of cast iron or steel and are designed for burning coal. Cast iron boilers are more resistant to corrosion than steel ones. Corrosive water can be improved with chemicals. Proper water treatment can greatly prolong the life of steel boiler tubes. Coal hand-fired and automatic heaters are able to replace electric, oil-fired, and gas boilers which are compact, self-contained units with a completely enclosing jacket.

Conventional radiators are set on the floor or mounted on the wall. The newer types may be recessed in the wall. Insulate behind recessed radiators with one-inch insulation board, a sheet of reflective insulation, or both.

Radiators may be partially or fully enclosed in a cabinet. A full cabinet must have openings at top and bottom for air circulation. The preferred location for radiators is under a window.

Baseboard radiators are hollow or finned units that resemble and replace the conventional wood baseboard along outside walls. They will heat a well-insulated room uniformly, with little temperature difference between floor and ceiling.

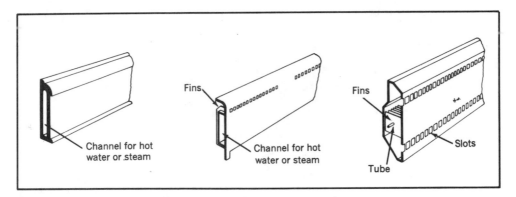

Fig. 60

Convectors usually consist of finned tubes enclosed in a cabinet with openings at the top and bottom. Hot water or steam circulates through the tubes. Air comes in at the bottom of the cabinet, is heated by the tubes, and goes out the top. Some units have fans for forced-air circulation. With this type of convector, summer cooling may be provided by adding a chiller and the necessary controls to the system. Convectors are installed against an outside wall or recessed in the wall.

Forced-Hot-Water Heating

Forced-hot-water heating systems are recommended over the less efficient gravity-hot-water heating systems.

In a forced-hot-water system, a small booster or circulating pump forces or circulates the hot water through the pipes to the room radiators or convectors.

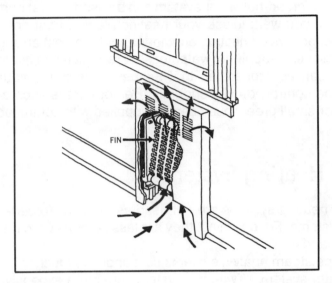

Fig. 61

Two-pipe forced-hot-water systems have two supply pipes or mains. One supplies the hot water to the room heating units, and the other returns the cooled water to the boiler.

In a one-pipe system, one pipe or main serves for both supply and return. It makes a complete circuit from the boiler and back again. Two risers extend from the main to each room heating unit. A two-pipe system has two pipes or mains. One carries the heated water to the room heating units; the other returns the cooled water to the boiler.

A one-pipe system, as the name indicates, takes less pipe than a two-pipe system. However, in the one-pipe system, cooled water from each radiator mixes with the hot water flowing through the main, and each succeeding radiator receives cooler water. Allowance must be made for this in sizing the radiators—larger ones may be required further along in the system.

Because water expands when heated, an expansion tank must be provided in the system. In an "open system," the tank is located above the highest point in the system and has an overflow pipe extending through the roof. In a "closed system," the tank is placed anywhere in the system, usually near the boiler. Half of the tank is filled with air, which compresses when the heated water expands. Higher water pressure can be used in a closed system than in an open one. Higher pressure raises the boiling point of the water. Higher temperatures can therefore be maintained without steam in the radiators, and smaller radiators can be used. There is almost no difference in fuel requirements.

With heating coils installed in the boiler or in a water heater connected to the boiler, a forced-hot-water system can be used to heat domestic water year-round. If you want to use your heating plant to heat domestic water, consult an experienced heating engineer about the best arrangement.

One boiler can supply hot water for several circulation heating systems. The house can be "zoned" so that temperatures of individual rooms or areas can be controlled independently. Remote areas such as a garage, workshop, or small greenhouse can be supplied with controlled heat.

Central Heating Systems

Steam heating systems are not used as much as forced-hot-water or warm-air systems. For one thing, they are less responsive to rapid changes in heat demands.

One pipe steam heating systems cost about as much to install as one-pipe hot-water systems. Two-pipe systems are more expensive.

The coal heating plant must be below the lowest room heating unit unless a pump is used to return the condensate to the boiler.

Cool Range Boiler

Coal-fired boilers for hot-water heating are compact and are designed for installation in a closet, utility room, or similar space, on the first floor if desired. In the United Kingdom, for example, solid-fuel gravity-feed central heating boilers are relatively easy to operate and simple to service. Typically, English range boilers may be installed in kitchens, utility rooms, or basement and are suitable for use with modern pumped systems (or existing traditional European gravity systems). Solid-fuel gravity-feed central boilers are generally automatic systems that keep the burning rate under close control and can be used with normal room thermostat and time clock. Range boilers have control thermostat, fan unit, and manual or automatic clinker ejection assembly, and are designed primarily to burn anthracite fuel.

The coal gravity-feed central heating boilers (45,000 to 60,000 Btu/hr.) use an integral water-sensing thermostat that switches an "excess air" fan

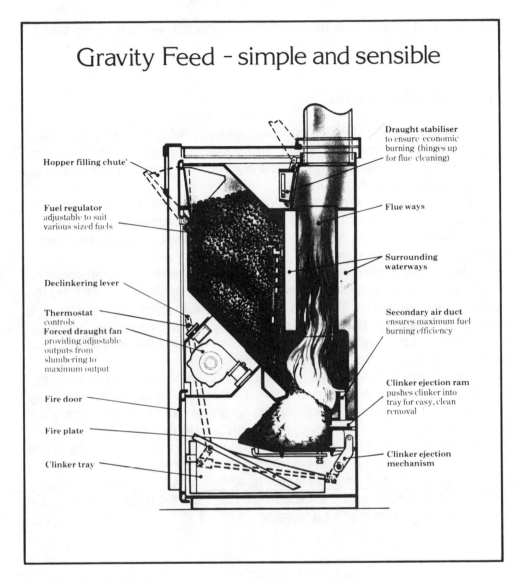

Gravity Feed - simple and sensible

Hopper filling chute'

Fuel regulator adjustable to suit various sized fuels

Declinkering lever

Thermostat controls
Forced draught fan providing adjustable outputs from slumbering to maximum output

Fire door

Fire plate

Clinker tray

Draught stabiliser to ensure economic burning (hinges up for flue cleaning)

Flue ways

Surrounding waterways

Secondary air duct ensures maximum fuel burning efficiency

Clinker ejection ram pushes clinker into tray for easy, clean removal

Clinker ejection mechanism

Fig. 62—Gravity feed

on to boost the fire when the room thermostat demands heat (or additional hot water is required for washing or other household activities) and, when the required temperature is achieved, it automatically switches the fan off to allow the fire to slumber economically under controlled natural draft until the next demand occurs. As the solid pea coal is consumed, the firebed is automatically replenished by gravity from a large hopper above (46 pounds to 61 pounds of coal), and a constant firebed depth is maintained at all times. This procedure insures constant optimum combustion efficiency which is further controlled by an inbuilt draft stabilizer. Residual ash forms into compact, solid clinker units which can then be removed by the ejection system, i.e., a simple dust-free method involving but two pulls of the ejector handle.

Automatic Boiler Controls

Each type of coal heating plant requires special features in its control system. But even the simplest control system should include high-limit controls to prevent overheating. Limit controls are usually recommended by the equipment manufacturer.

The high-limit control, which is usually a furnace or boiler thermostat, shuts down the fire before the furnace or boiler becomes dangerously or wastefully hot. In steam systems, it responds to pressure; in other systems, it responds to temperature.

The high-limit control is often combined with the fan or pump controls. In a forced-hot-water system, these controls are usually set to start the fan or the pump circulating when the furnace or boiler warms up and to stop it when the heating plant cools down. They are ordinarily set just high enough to insure heating without overshooting the desired temperature and can be adjusted to suit weather conditions.

Other controls insure that all operations take place in the right order. Room thermostats control the burner or stoker on forced systems. They are sometimes equipped with timing devices that can be automatically set to change the temperatures desired at night and in the daytime.

Since the thermostat controls the house temperature, it must be in the right place—usually on an inside wall. Do not put it near a door to the outside, at the foot of an open stairway, above a heat register, television, or lamp, or where it will be affected by direct heat from the sun. Check it with a good thermometer for accuracy.

Stoker-Fired Coal-Burner Controls

The control system for a coal stoker is much like that for an oil burner. However, an automatic timer is usually included to operate the stoker for a few minutes every hour or half hour to keep the fire alive during cool weather when little heat is required.

A stack thermostat is not always used, but in communities where electric power failures may be long enough to let the fire go out, a stack thermostat or other control device is needed to keep the stoker from filling the cold firepot with coal when the electricity comes on again. Sometimes a light-sensitive electronic device such as an electric eye is used. In the stoker-control setup for a forced-warm-air system, the furnace thermostat acts as high-limit and fan control.

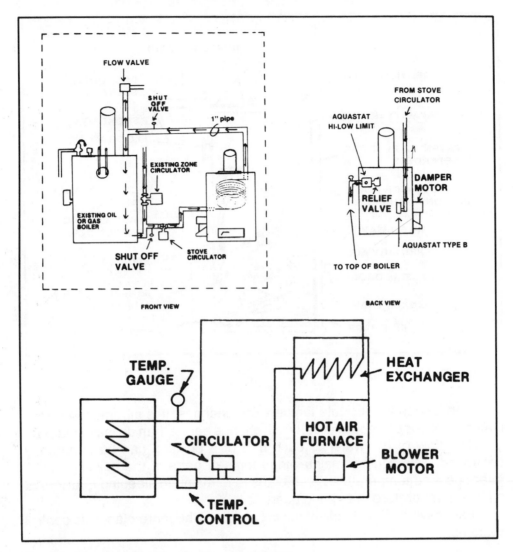

Fig. 63—Energen Heating System Model 101

Hot-water or steam heat distribution systems may be controlled in other ways. If the furnace or boiler heats domestic water, more controls are needed.

In some installations of forced-hot-water systems, especially with domestic-water hookups, a mixing valve is used. The water temperature of the boiler is maintained at some high, fixed value, such as 200° F. Only a portion of this high-temperature water is circulated through the heating system. Some of the water flowing through the radiators bypasses the boiler. The amount of hot water admitted is controlled by a differential thermostat operating on the difference between outdoor and indoor temperatures. This installation is more expensive than the more commonly used control systems, but it responds almost immediately to demands; and,

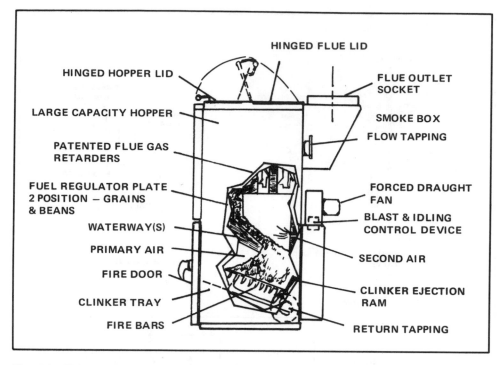

Fig. 64—Trianco Boiler

although it cannot anticipate temperature changes, it is in a measure regulated by outside temperatures, which change earlier than do those indoors.

The flow of hot water to each part of a building can be separately controlled. This zoning (maintaining rooms or parts of the building at different desired temperatures) can be used to maintain sleeping quarters at a lower temperature than living quarters.

Fuel savings help to offset the initial cost of the more elaborate control systems.

Hand-Fired Boilers

Many modern homes were designed for the convenience of central heating. The layout of the house, with many small, scattered rooms sealed off from each other by walls, closets, and other heat-blocking barriers, may make heating with stoves virtually impossible. All-electric homes are often especially difficult to heat with stoves. This type of layout is ideal for a wood or multi-fuel furnace. Coal boilers are available that will heat any size home and are designed for using hot water. In most cases the boiler is designed to replace the existing boiler using the pipes currently in place.

Maintenance of such a unit is more demanding than for a conventional oil burner. The fire must be fed, the ash box cleaned out, and they burn

Fig. 65—The Mascot MC-200 boiler can be used with gravity or forced circulation systems. Options in tandem with another boiler are gravity, cold start-forced circulation and summer/winter hook-ups.

large amounts of fuel. Typical burning times are 12 hours or more. They may not be legal in some states, as suitable code requirements have not been developed to ensure safe construction and maintenance. Totally automatic systems are also available.

The one prerequisite to a coal boiler is a large supply of cheap fuel. The average boiler uses tons of coal per year. A multi-fuel furnace can use any fuel source with oil or gas acting as a backup if the alternative fuel burns out. The boilers are thermostatically controlled. When no heat is required the damper is closed off so the fuel barely burns; when heat is required, the dampers open and the fire comes back to life.

Selected Illustrated Boilers

The Franco-Belge[1] boilers for central heating use are specially designed to produce a high output and to burn coal. The boiler has a gravity-feed type of hearth so that it can be stoked with a large amount of coal.

[1]Franco-Belge Foundries of America Inc. 70 Pine Street, New York, New York 10005

Fig. 66

The heat output is efficiently transmitted to the circulating central heating water. The rate of burning is thermostatically controlled from the temperature of central heating water.

There is a choice of three central heating boilers; each of these units has a large hot plate giving a very useful extra cooking facility. The Franco-Belge takes up a minimum of floor space and provides domestic hot water and full central heating through as many as 15 radiators.

The operating principles of the Franco-Belge boilers are easy to understand. The thermostat regulates the burning rate to ensure a constant predetermined central heating water temperature. If the water temperature falls, the thermostat opens an air inlet which boosts the fire. Similarly, as the water temperature rises, the inlet closes to damp down the fire to maintain a roughly constant water temperature.

The draft slide in the firebox door introduces additional air for lighting the fire, boosting the output after slow night-burning, and providing extra heat when using the oven and hot-plates for cooking. Thus, it is possible to maintain the heat output into the central heating, as well as provide additional heat for cooking and baking.

The firebox is designed to use all solid fuel: coal, peat, or smokeless fuels. Complete combustion of all fuels is ensured by secondary air inlets positioned at the top of the firebox which introduce extra air to burn off all volatile gases.

A highly sensitive control automatically regulates the air inlet and thus the burning rate.

The heat exchanger is made of special steel and provides a large surface area. This produces a highly efficient heating appliance.

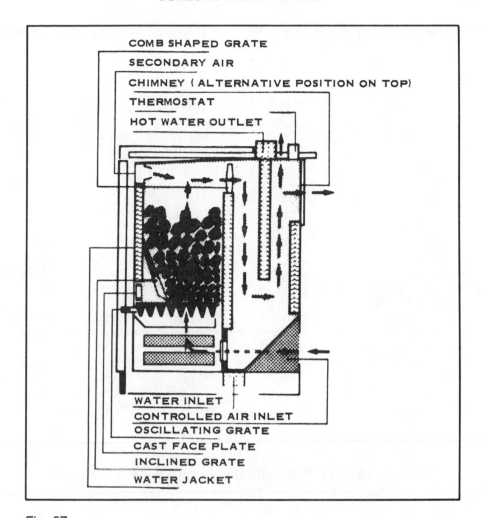

COMB SHAPED GRATE
SECONDARY AIR
CHIMNEY (ALTERNATIVE POSITION ON TOP)
THERMOSTAT
HOT WATER OUTLET

WATER INLET
CONTROLLED AIR INLET
OSCILLATING GRATE
CAST FACE PLATE
INCLINED GRATE
WATER JACKET

Fig. 67

The heat exchanger and the flue passages can all be easily cleaned through special access covers.

Because of its original design, the Franco-Belge central heating cooker meets the requirements of both a cooker and a central heating boiler. Its high efficiency makes it an especially economical appliance. In any installation, relevant building codes and practices must be observed. The appliance is not designed as a pressure vessel, an important factor being that the circuit is left open to the atmosphere and is in no way constructed as to allow any pressure build-up to occur. The provision of an A.S.M.E. temperature/pressure relief valve near the cooker is essential. A gravity circuit MUST be provided, as a fail-safe heat loss in the event of a circulating pump failure. The advantages of this will be realized when the heating is maintained in times of power-out. To achieve this, ensure that large-diameter pipes leading to upstairs radiators have a direct flow from the boiler, or install a big hot water tank with large-diameter heat-exchanging coil, situated above the cooker.

TABLE 11
FRANCO-BELGE BOILER TECHNICAL DETAILS

	190.05	190.09	190.13
Btu /Hr.	35,000	52,000	70,000
Dimensions of Boiler			
Width	13 ins	17 ins	19½ ins
Depth	25 ins	29½ ins	32 ins
Height	31½ ins	31½ ins	31½ ins
Firebox Size			
Width	10 ins	14 ins	16½ ins
Depth	10 ins	12½ ins	14 ins
Height	13 ins	12 ins	12 ins
Diameter of Chimney Outlet	6 ins	6 ins	6 ins
Distance from Floor to Center			
of Rear Chimney Outlet	26½ ins	26½ ins	26½ ins
Outlet Water Pipe Thread	1½ ins	1½ ins	1½ ins
Inlet Water Pipe Thread	1½ ins	1½ ins	1½ ins
Distance from Floor to Water			
Inlet Connection	4½ ins	4½ ins	4½ ins
Capacity of Water Jacket	5½ gals	7 gals	8 gals
Weight Packed	320 lbs	410 lbs	500 lbs

TABLE 12
APPROXIMATE DIMENSION IN INCHES (SEE FIG. 68)

BOILER TYPE	A	B	C	D	E	F
190–05	4.0	6.8	3.6	9.8	5.0	5.2
190–09	6.0	8.1	5.6	9.8	8.6	5.6
190–13	7.1	8.5	6.75	9.8	10.0	5.6

The Fawcett Compact Coal Furnace[2] was designed to satisfy the demand for a smaller heating unit capable of looking after the requirements for today's modern-sized, well-insulated homes.

Model CWF 85, although similar in size and output to the WF 200 model, offers the same quality features: it is of very efficient, airtight construction and thermostatically controlled. The combustion chamber is constructed of one-eighth-inch steel plate, lined with firebrick, and is equipped with cast iron grates, a necessary requirement for burning coal.

[2]Fawcett Enheat Ltd., Sackville, New Brunswick, N.S., Canada EoA3CO

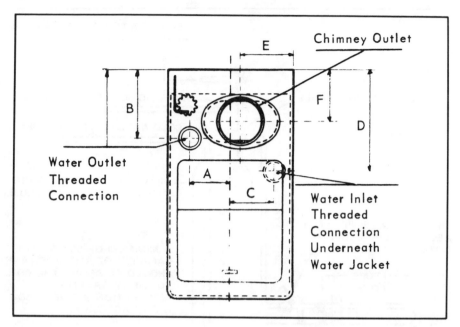

Fig. 68

The heat exchanger is constructed of 16-gauge steel and features a dual heat baffle to assist in achieving maximum heat output.

The easily removed fan box can also be installed on either side of the furnace if desired.

During the heating season, the Fawcett add-on will operate as a primary heating unit, giving all-night heat from a full charge of fuel. When there is no one at home to tend the boiler, your existing furnace will automatically take over from the add-on, providing uninterrupted heating.

In power failure the system can be operated on gravity, eliminating the fear of pipes freezing. In most cases a tie-in system can be installed in less than a day with little interruption to heating service.

Thermo-control efficiency with the convenience of coal is offered by National Stove Works[3] in the Thermo-Control Model 550, the "ultimate coal-burning system."

Overall size of the Model 550 Coal Stove is the same as the Thermo-Control Model 500. Like the 500, this stove can be used as a radiant heater, to supply domestic hot water, or can be attached to a central heating system—either forced-warm-air or hot-water heat.

The Thermo-Control Model 550 features a coal combustion chamber equipped with a rotary shaker grate and separate ash pan. This rugged unit accommodates 50 pounds of coal and easily sustains a fire overnight. Its steady heat compares favorably to the output of a gas or oil boiler rated at 85,000 Btu.

---

[3]National Stove Works, Inc., Cobleskill, N.Y. 12043

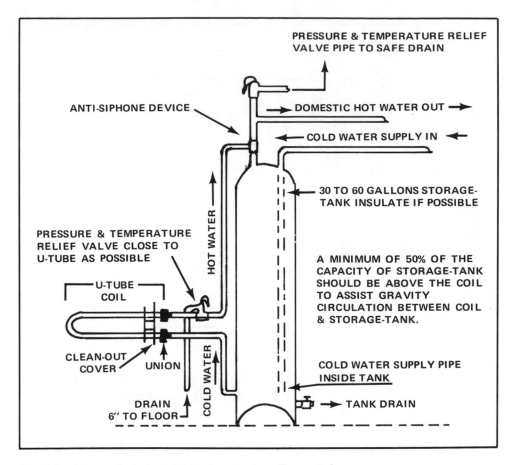

PRESSURE & TEMPERATURE RELIEF
VALVE PIPE TO SAFE DRAIN

ANTI-SIPHONE DEVICE

DOMESTIC HOT WATER OUT →

COLD WATER SUPPLY IN ←

30 TO 60 GALLONS STORAGE-
TANK INSULATE IF POSSIBLE

HOT WATER

PRESSURE & TEMPERATURE
RELIEF VALVE CLOSE TO
U-TUBE AS POSSIBLE

A MINIMUM OF 50% OF THE
CAPACITY OF STORAGE-TANK
SHOULD BE ABOVE THE COIL
TO ASSIST GRAVITY
CIRCULATION BETWEEN COIL
& STORAGE-TANK.

U-TUBE
COIL

CLEAN-OUT
COVER   UNION

COLD WATER SUPPLY PIPE
INSIDE TANK

COLD WATER

DRAIN
6" TO FLOOR

TANK DRAIN

Fig. 69—Domestic hot water coil option on Fawcett furnaces

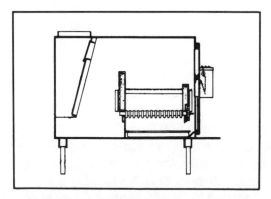

Fig. 70—Cutaway diagram

The Model 550 delivers high efficiency with virtually complete coal combustion, thanks to its unique design and strategic provision for combustion air. Wood may also be burned in it if desired.

The Thermo-Control Model 550 measures 33 inches high × 24 inches wide and 40 inches long, weighs 500 pounds, and has an eight-inch diameter flue. Heating range is up to 2,500 square feet.

The hook-up illustrated in the accompanying figure demonstrates how simply the Thermo-Control domestic hot water model can be installed. A three-quarter-inch line is run from the bottom outlet of the existing water heater or storage tank to the bottom of a Thermo-Control side coil. A three-quarter-inch line is then run from the top of the hot water tank. With proper fittings and controls installed as shown the wood stove will safely heat the water from the tank as it passes through the pipe loop by thermo-convection. Should the water temperature in the storage tank drop below the temperature setting, the existing water heater will act as a backup unit by automatically "switching on" and bringing the water up to temperature.

Following are the guidelines for this simple installation:

The Thermo-Control Domestic Hot Water Model Stove must always be installed with a storage tank or storage-type hot water heater (electric/gas or oil-fired).

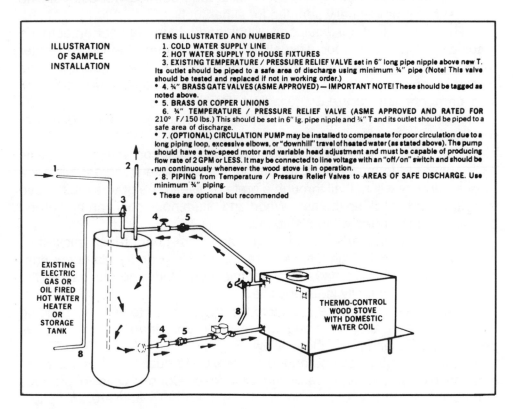

Fig. 71—Thermo Control 550. In the event of power failure (especially when circulating pump is required) the thermostatic control handle on the stove should be kept at or lowered to a setting which will prevent discharge at the temperature/pressure relief valve.

The storage tank or existing water heater should have a minimum storage capacity of 30 gallons and should be well constructed and insulated to prevent corrosion and heat loss. It must be rated for a minimum of 150 pounds-per-square-inch working pressure.

All pipes, tubing, and fittings should be non-corrosive and of minimum three-quarter-inch inside diameter.

Relief valve outlets should be piped to safe areas of discharge using minimum three-quarter-inch pipe.

Only gate valves should be used—do not use globe valves or any other type of "complete-shutoff" type valves.

Locate the stove as close to the existing water heater as is possible, leaving enough room to service the system safely. Keep length of pipe runs and number of elbows and fittings at a minimum.

The Energy Marketing Home Heater[4] is a true combination coal- and wood-burning unit. It was carefully engineered initially to burn both anthracite (hard) and bituminous (soft) coal. Then the stove was expanded upward to create the large firebox necessary for a long-burning woodstove. The end result is a unit equally adapted for either fuel.

The Home Heater is equipped with a manual turn damper on both the loading door and ash door. This allows the air entering the Home Heater to be properly proportioned both above and below the fire to guarantee the efficient burning of both the fuel and the volatile gases arising from the fuel. The dampers are very easy to adjust. Being able to see through their openings into the firebox and ash pit, the user is able to see the results of any adjustments made.

The Home Heater is equipped with heavy cast iron shaker grates as it is impossible to burn anthracite without them. Unlike wood ash, the ash from anthracite will not fall through a grate by itself. The grates must move to agitate and break up the ash. Wood and bituminous coal can be burned on these same grates without shaking.

The Home Heater's basic hull is made from quarter-inch-thick steel and is fully lined with high-temperature firebrick. The doors, dampers handles, and grates are all made of heavy cast iron. Under normal usage, the Home Heater is virtually indestructible.

Many accessories are available for the Home Heater. A "building block" approach allows starting with a basic unit, then "adding on" as personal needs and pocketbook allow.

A hot-air bonnet can be added to hook up to ductwork. Later a combination blower and filter pack can be added. Copper coils to produce domestic hot water and/or to tie into baseboard hot water are an option. A thermostatically controlled motorized damper for automatic operation is another option.

---

[4]Energy Marketing Corporation, P. O. Box 636, Bennington, VT 05201

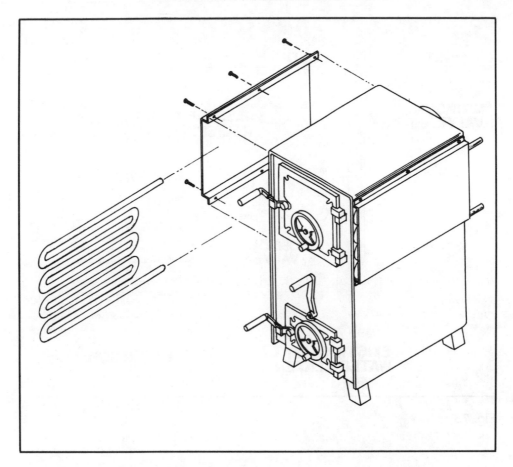

Fig. 72—Home Heater water coil Model HC1

The water coils for the Home Heater can be mounted on either side, on the top, or on all three surfaces. This option allows you to size your hot water production to your family's needs.

The coils are made from one continuous length of three-quarter-inch copper tubing. The tubing you connect to your coils should also be a minimum of three-quarter-inch diameter.

External coils are used with the Home Heater for three reasons:

1. No space inside the firebox is used, thus allowing the maximum amount of fuel to be loaded.

2. Being outside the stove allows the use of copper tubing. With copper tubing there is no rust problem, and no worry about burning the coils out if there is no water in them.

3. Copper transfers heat approximately 15 times as fast as steel. Thus much more heat can be transferred to your hot water and it can be done in a smaller space than with steel.

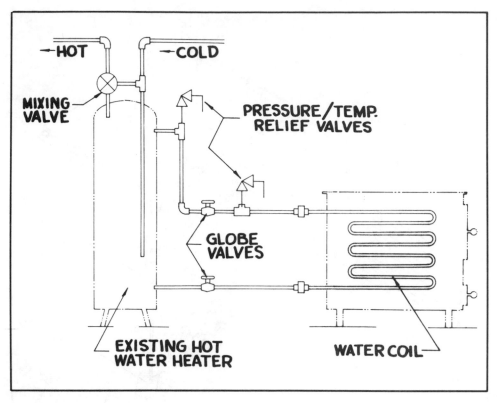

Fig. 73

Other Factors in Hot Water Heating

Gravity: You can get adequate circulation in a domestic system through gravity convection (a thermosyphon) if you observe the following:

1. Use large diameter pipe, preferably three-quarter-inch or larger.
2. Minimize all fittings, elbows, etc. Use gate valves where required.
3. Locate the storage tank as close as possible to the stove.

Use either a horizontal overhead tank directly above the stove or a vertical tank beside it. In either case, the hot water should come in at the top and colder water should leave at the bottom. The thermosyphon is dependent on having two vertical sections, one filled with hot water and the other with relatively colder water. The taller those sections are, the greater the lift. The overhead tank will produce lift but has more friction to overcome. Either system works.

Circulator: If you can't utilize gravity convection you'll have to use a circulator. In a domestic system the circulator should be bronze, not cast iron (again to avoid rust), and should be in the one- to three-gallon-per-minute range. The smallest circulator you can find is probably adequate to keep sufficient water moving through your coils.

Pressure Relief: All hot water systems must be safeguarded against excess pressure. If you're tying into an existing system it should already have adequate pressure relief valves (PRV). If your domestic system has a check valve or pressure regulator in the cold water inlet line, then it should have a pressure relief valve. (If there is no check valve, then pressure will merely back up into the feed line.) The PRV should be set above your static line pressure but below the safe working pressure of the weakest component in the system. This will generally mean a setting between 100 and 150 psi.

Temperature Relief: Domestic systems should be safeguarded against excess temperature. The temperature of the water in the holding tank will not be controlled directly but will vary according to operating conditions. You should first try to adjust the size of your coil so that under normal conditions the temperature in the tank does not become excessive. But just in case it does you should install temperature reliefs in the outlet line. These are intended to protect your family against excess temperature. (The system needs protection against excess pressure.)

You can use inexpensive fusible plugs, but once they open they have to be replaced to stop the flow. A better approach is a combination pressure-temperature relief valve (PTRV). These are available in the 75–100 psi range and open at 200–210° F. More importantly, they reset themselves automatically. A third approach is to install a mixing valve in the hot water outlet which automatically adds cold water to the existing hot water so as to achieve a desired temperature (adjustable) at the faucet. A mixing valve followed by a temperature relief valve or plug is the most satisfactory combination.

Tanks (domestic hot water): If you have an existing tank you'll want to utilize that. Otherwise, there are commercial holding tanks (without heat sources) available. They should be lined (rust again) and insulated. Make certain they have adequate fittings (large enough and in the right places) for the intended use. Tanks in the 30 to 50 gallon range are adequate.

Whatever system you attempt to set up, you should attempt it first with gravity convection. If you have to go to a circulator, then at least when the power goes off you'll get some circulation through gravity, hopefully enough to prevent a "blow-off." Make certain all your pressure and temperature reliefs are functioning properly. If in doubt, replace them or add additional ones. Learning to heat with coal or wood is going to involve some trial and error. As long as the errors only cost you a little water down the drain, you should be happy with the results.

Selected Automatic Coal-Fired Boilers and Stokers

The limitations of hand-fired heating equipment are not new. Constant feeding of the coal bed and mechanized stoking were of much interest at the turn of the century. Today hand-fired stoves and furnaces have automatic stokers, and others use large and small types of hoppers in order to increase the self-tending aspects of coal heating. There are several fully automatic coal boilers and furnaces available; the following material illustrates a few of the choices open to the modern homeowner.

The Axeman-Anderson Anthratube's[5] fuel bed is arranged so that anthracite is almost completely burned with an absolute minimum of excess air. Roughly, the average ton of anthracite contains 240 pounds of useless ash. Of the remaining 1760 pounds of usable fuel, when burned in the Axeman-Anderson Anthratube, 60 pounds go out with the ash, 200 pounds go up the chimney as wasted heat and 1500 pounds heat the house. On the average coal-fired job, of the remaining 1760 pounds of usable coal, 60 pounds go out with the ash, 800 pounds go up the chimney as waste heat, and 900 pounds heat the house.

The room thermostat controls the temperature. When heat is required, the motor which operates the induced draft fan, coal tube, and reciprocating grate is automatically started.

Coal flows up the spiral path formed by helicoidal ribbon welded inside the coal tube. Coal actually "falls uphill" and flows only when the coal tube rotates. The coal tube cannot overfeed and cannot jam like the conventional screw-feed coal conveyor. This is a completely new method of feeding coal.

Coal gets only as high in the transfer head as shown. No excess coal is delivered because coal merely slips back through the hollow center when there's enough in the transfer head.

The grate (with a lifetime guarantee) is made of heavy solid steel (one-half inch and three-quarters of an inch thick) of stepped type moving two inches forward and two inches to the rear. It is quiet and simple in operation.

A. The weight of the fire bed forces ash onto the grate.

B. As the grate moves slowly forward, steps on the grate cut ash from the bottom of the fire bed and push it toward the container.

C. As the grate moves slowly back, ash at the end of the grate falls off into the container or pit. The grate operates only when the fan is running which accounts for clean operation.

---

[5]Axeman-Anderson Company, 233 West Street, Williamsport, PA 17701

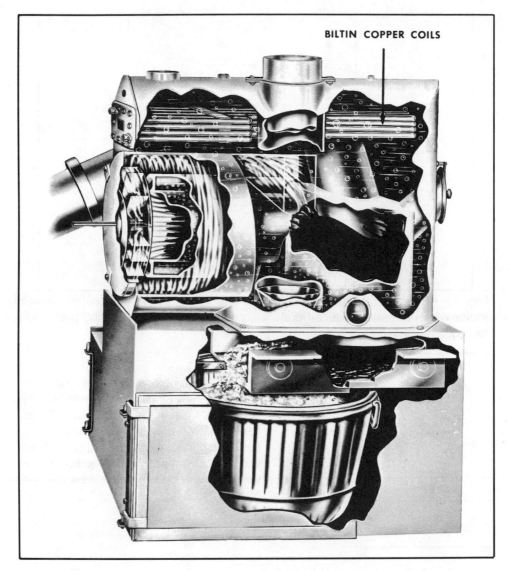

BILTIN COPPER COILS

Fig. 74—The Axeman-Anderson Anthratube's fuel bed is arranged so that anthracite is almost completely burned with an absolute minimum of excess air.

The flue gases are whirled at great velocity in the centrifugal heat
absorber. This repeated whirling with its intense scrubbing action results
in a very high rate of heat transfer from the flue gases to the boiler water.

Fly ash always results from burning anthracite at high rates. In the
Axeman-Anderson Anthratube it is whirled through the centrifugal heat
absorber along with the gases, creating a self-cleaning action. At the end
of the centrifugal heat absorber, the whirling gases and fly ash are led into
a special adaptation of the cyclone separator. This cyclone separator not
only further absorbs heat from the flue gases but it also extracts nearly all

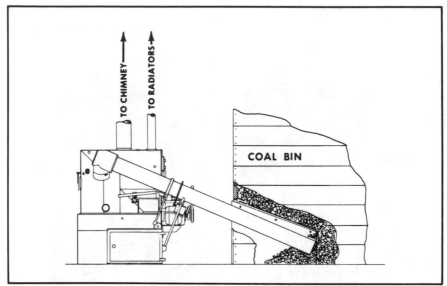

Fig. 75—The room thermostat controls the temperature. When heat is required, motor which operates induced draft fan, coal tube and reciprocating grate is automatically started.

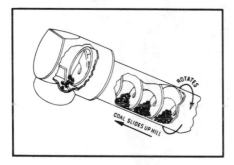

Fig. 76—Coal flows up the spiral path formed by helicoidal ribbon welded inside coal tube. Coal actually "falls uphill" and flows only when coal tube rotates. The coal tube cannot overfeed and cannot jam like the conventional screw feed coal conveyor. This is a completely new method of feeding coal.

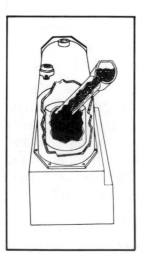

Fig. 77—Coal gets only as high in transfer head as shown. No excess coal is delivered because coal merely slips back through hollow center when there's enough in the transfer head.

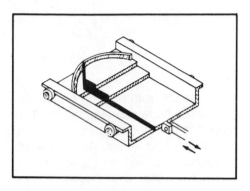

Fig. 78—The grate with a lifetime guarantee is made of heavy solid steel (½″ and ¾″ thick) of stepped type moving 2″ forward and 2″ to the rear. Quiet and simple in operation.

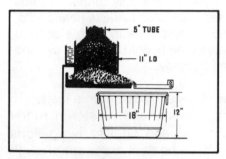

Fig. 79 A—Weight of fire bed forces ash onto grate.

Fig. 79 B—As grate moves slowly forward, steps on grate cut ash from bottom of fire bed and push it toward container.

Fig. 79 C—As grate moves slowly back, ash at end of grate falls off into container or pit. Grate operates only when fan is running which accounts for clean operation.

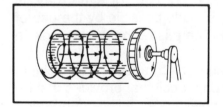

Fig. 80—The flue gases are whirled at great velocity in the centrifugal heat absorber. This repeated whirling with its intense scrubbing action results in a very high rate of heat transfer from the flue gases to the boiler water.

of the fly ash formed and drops it down through a cone to the grate. Flue gas leaves the upper part of the cyclone going from there to the chimney.

Hot boiler water surrounds copper coils, quickly transferring its heat through the copper walls to the water inside. As this domestic water receives heat, it flows instantly to the bath, laundry, or kitchen (in the case of the tankless or instantaneous type). A tempering valve ensures that this water

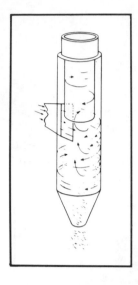

Fig. 81—Fly ash always results from burning anthracite at high rates. In the Axeman-Anderson Anthratube it is whirled through the centrifugal heat absorber along with the gases, creating a self-cleaning action. At the end of the centrifugal heat absorber, the whirling gases and fly ash are led into a special adaptation of the cyclone separator. This cyclone separator not only further absorbs heat from the flue gases but it also extracts nearly all of the fly ash formed and drops it down through a cone to the grate. Flue gas leaves the upper part of the cyclone going from there to the chimney.

is always at approximately 140°F. When a storage-type heater is installed, water that is heated flows rapidly to a storage tank where it is stored, ready for instant use. The tankless heaters supply three and a half to five gallons of piping hot water per minute. The storage type heaters are large enough to handle a 70- to 150-gallon storage tank. Since domestic water comes in contact only with copper and bronze in the heater, it's always clean and rustless.

Keystoker[6] units include boilers and warm air furnaces. There is a simple feed mechanism with very few moving parts which can easily be adjusted and maintained by the average housewife. Troublesome belts, shear pins, and worm conveyors have been eliminated. This feed mechanism with slow-moving, long-life parts keeps service calls at a minimum, the manufacturer asserts. Accurate controls and thermostat provide automatic, well-regulated warmth with quick response to weather changes. Warm air, water, or steam heat are equally efficient with the Keystoker unit. Abundant hot water, 200 gallons or more, every hour of every day, is available. This year-round plentiful supply is made possible by the use of copper coils that heat your water in a jiffy. There is no hot water tank to install.

The Keystoker boiler includes: boiler gauge; high capacity steam chest, for a large supply of steam or hot water; copper coil that provides instantaneous hot water; limit control that keeps boiler water at "just right" temperature; vertical water baffle that increase internal heat recovery action; downdraft multipass steel boiler—entirely surrounded by water walls forcing hot vapors to brush the entire inner surface—thus extracting the maximum amount of useful heat, and feed mechanism—easily and quickly adjusted. Also, the entire boiler is made of heavy-gauge, high-quality steel.

---

[6]Keystone Manufacturing Co., Schuylkill Haven, PA.

The hot water specifications for Keystoker coal stoker boiler are shown in Table 13.

TABLE 13

MODEL NO.	Total Capacity Square Feet		Recommended Load Square Feet		Hopper Capacity	STOKER DATA			
	Steam	Hot W.	Steam	Hot W.	Lbs.	Model	No. of Grates 3" × 14"	Lbs. Per Hr.	Smoke Outlet Size
KA-4	400	640	280	450	225	A	3	20	8"
KA-6	600	960	420	675	275	A	4	30	9"
KB-8	800	1280	560	900	315	B	4	30	9"
KC-10	1000	1600	700	1125	450	C	5	40	10"
KD-12	1200	1920	840	1350	525	D	5	40	10"
KE-15	1500	2400	1050	1690	525	E	7	60	12"
KF-18	1800	2880	1260	2025	525	F	8	70	12"
KG-22	2200	3520	1540	2475	585	G	9	80	14"

The Losch[7] water-cooled stokers were developed during the mid 1930's. These units were considered among the best of anthracite technology. They could accept a wide range of ash fusion temperatures and sizes and were available in a wide range of ratings. By fall of 1977, several units were available for use in fire-tubed or water-tubed boilers. These include:

Model 24S—24 inches wide with twelve-inch-long burning area. Rated 720,000 Btu/hr. with 70 pounds of coal per hour feed.

Model 96L—4,896,000 Btu/hr. with 490 pounds per hour of coal.

These units take rice or buckwheat with various ash fusion temperatures.

Van Wert Anthraterm[8] boilers for residential and small commercial applications are available.

TABLE 14

MODEL #	COAL FEED LBS./HR.
600	18
800	24
1200	34
1500	40
1800	50
2400	60
4000	100

---

[7]Losch Boiler Company, Summit Station, Schuylkill County, PA.
[8]Van Wert Manufacturing Company, 739 River Street, Peekville, PA 18452.

Anthracite coal is fed automatically from the bin, and ashes are dropped in a steel basket. The firebed of live coals under constant thermostatic control prevents the needless waste of over-firing. Van Wert units are available for hot water, hot air, and steam.

Several sizes of anthracite coal can be burned in the Electric Furnace-Man[9] Model 520 anthracite-fired stoker equipment. Another feature of this equipment is the ability to burn anthracite with an ash fusion temperature as low as 2500°F. According to its manufacturers, if properly lubricated and cleaned once a year, the system requires minimal replacement parts during the first twenty years of installation. This statement, they claim, is based on empirical data. An important part of the successful operation of these stokers is the proper design of coal-handling facilities within the structure. Although coal feeding can be an automatic operation, ash handling remains manual with attention required several times per week depending on the severity of the weather. The units operate satisfactorily for summer domestic-water-heating loads with a fringe benefit of keeping the flue and chimney above dew point, thereby reducing corrosion.

Electric Furnace-Man stoker boiler burner units for hot water or steam follow the latest engineering practice and are built to ASME code requirements and bear the code seal of approval. Boilers are made of extra heavy steel plate (not less than one-quarter inch throughout) with the best quality workmanship and materials, and utilize the plate flue principle. Heating surfaces are in a vertical position, which assures maximum heating efficiency—faster heat transfer and lower fuel consumption. The use of baffles insures more thorough absorption of the heat for useful purposes—prevents its escaping up the stack and being wasted. Right through the heating season you will have a uniformly clean unit. This insures constant high heating efficiency.

Electric Furnace-Man is completely automatic, simple in design, and all moving parts are accurately machined for quiet operation. Coal is automatically moved from bin to the burner and ash if deposited in a container in the dust-tight base. The burner with its clean, bright, smokeless, intensely hot flame is controlled automatically by the thermostat. The burner mechanism can be located on either side of the AP-350 and 520 with the bin feed from the opposite side, and is supplied with revolving retort and automatic fine clean-out. Bin feed can be placed in any one of six positions. "You can leave for days at a time, secure in the knowledge that your home will be warm, clean and comfortable when you return," the company asserts.

Will-Burt domestic stokers[10] (hopper and bin feed) are *not* for anthracite coal. There are several models, Numbers S–20, S–30, S–50, and S–75.

---

[9]Electric Furnace-Man, Division of General Machine Co., Inc., Emmaus, PA 18049
[10]The Will-Burt Company, 169 S. Main Street, Orrville, OH 44667.

TABLE 15

	20	30	50	75
STOKER NUMBER	20	30	50	75
COAL BURNING RATE (Approximate Lbs. Per Hr.	12	14	25	30
Continuous Operation)	15	20	37	53
	20	30	50	75
GROSS OUTPUT 1,000 Btu per hour	156	234	390	585
BOILER H.P.	4.7	7.0	11.6	17.4
MOTOR SIZE—H.P. (Hopper and Bin Feed Models)	¼	¼	⅓	⅓
HOPPER CAPACITY	350	350	600	600
COAL FEED TO BE—Dia. (Inches)	3	3	3	3
WEIGHT CRATED (Approximate Lbs.)				
HOPPER MODEL—Refractory	350	350	400	450
BIN FEED MODEL—Refractory		400	450	550

According to the manufacturers, Will-Burt stoker-fired units develop heat by the safest and most economical method. That is, automatically controlled temperatures within the furnace as well as throughout the home eliminate fire hazards and prolong furnace life. Stoker heat is clean; chimney smoke is eliminated. There are no grates to shake or dusty ashes to handle. This coal burner is "bin fed"; it delivers the coal from your bin right into the furnace automatically. All joints between the bin and the furnace are well sealed. The continuous gear-driven transmission is rugged in construction and has been tried and proved over a period of many years. All domestic models are furnished with this standard transmission, single and shaft motors, and three-speed pulleys for easy coal-feed adjustment.

The J. E. Williamson Company[11] is an established foundry and machine shop that has supplied replacement parts for all types of anthracite- and bituminous-fired stokers. Several years ago the company engaged in the fabrication of the Williamson-McClave Mini Stoker. The units are too large for residential applications; however, a number of small commercial applications utilize this equipment. The publicized size range for this equipment is 30- to 500-boiler horsepower.

The anthramatic principle is the most modern method of automatically burning anthracite coal in the home. It is entirely new, simple and inexpensive. The anthramatic takes coal from the bin by worm, feeds it over the top of the grate, burns it efficiently and drops the ashes off the edge of the grate into the handy ash container. There is no shoveling of coal or ashes. Anthramatic heating has been built to the requirements of a boiler pressure vessel code of the American Society of Mechanical Engineers.

---

[11]J. E. Williamson Company, Inc., Bellwood, PA 16617

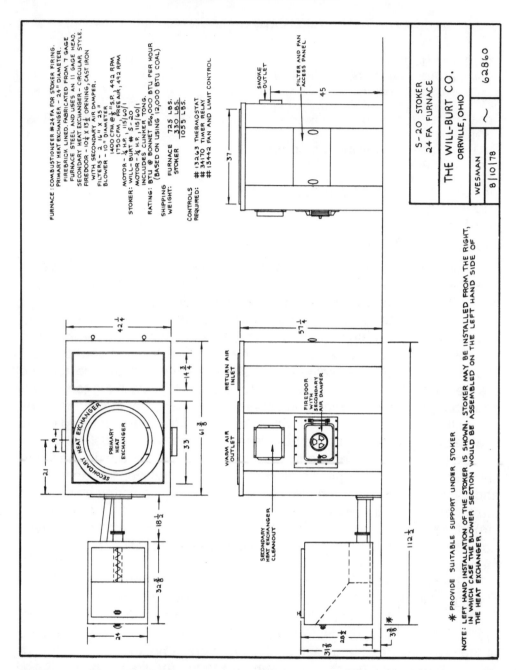

Fig. 82—S-20 Stoker 24FA Furnace by Will-Burt

Chapter 12

Warm Air Coal Furnaces

Two little dogs sat by the fire
Over a fender of coal dust;
Said one little dog to the other little dog,
"If you don't talk, why must I?"

Mother Goose

Forced-Warm-Air Heating

Forced-warm-air heating systems are more efficient and cost more to install than gravity-warm-air heating systems.

Forced-warm-air systems consist of a furnace, ducts, and registers. A blower in the furnace circulates the warm air to the various rooms through supply ducts and registers. Return grilles and ducts carry the cooled room air back to the furnace where it is reheated and recirculated. Most installations have a cold-air-return in each room (except the bathroom and kitchen). If the basement is heated, additional ducts should deliver hot air near the basement floor along the outside walls. In cold climates, a separate perimeter-loop heating system may be the best way to heat the basement.

Forced-warm-air systems heat uniformly and respond rapidly to changes in outdoor temperatures. They can be used in houses with or without basements—the furnace need not be below the rooms to be heated nor centrally located. Some can be adapted for summer cooling by the

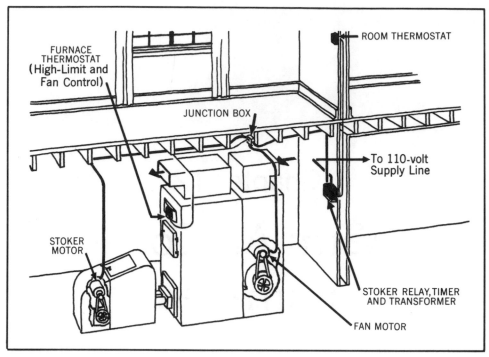

Fig. 83—Controls for a stoker-fired coal burner with a forced-warm-air heating system.

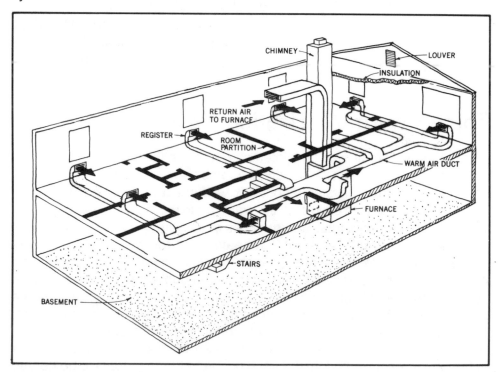

Fig. 84—Forced-warm-air systems are the most popular-type of heating systems.

addition of cooling coils. Combination heating and cooling systems may be installed. The same ducts can be used for both heating and cooling.

The warm air is usually filtered through inexpensive replaceable or washable filters. Electronic air cleaners can sometimes be installed in existing systems and are available on specially designed furnaces for new installations. These remove pollen, fine dust, and other irritants that pass through ordinary filters and may be better for persons with respiratory ailments. The more expensive units feature automatic washing and drying of the cleaner.

A humidifier may be added to the system to add moisture to the house air and avoid the discomfort and other disadvantages of a too-dry environment.

Warm-air-supply outlets are preferably located along outside walls. They should be low on the wall, in the baseboard, or in the floor where air cannot blow directly on room occupants. Floor registers tend to collect dust and trash, but may have to be used in installing a new system in an old house.

High-wall or ceiling outlets are sometimes used when the system is designed primarily for cooling. However, satisfactory cooling as well as heating can be obtained with low-wall or baseboard registers by increasing the air volume and velocity and by directing the flow properly.

Ceiling diffusers that discharge the air downward may cause drafts; those that discharge the air across the ceiling may cause smudging.

Most installations have a cold-air return in each room. When supply outlets are along outside walls, return grilles should be along inside walls

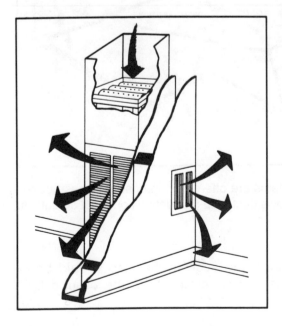

Fig. 85—Vertical furnaces installed in a closet or a wall recess or against the wall are popular in small houses. The counterflow type (shown here) discharges the warm air at the bottom.

in the baseboard or in the floor. When supply outlets are along inside walls, return grilles should be along outside walls.

Centrally located returns work satisfactorily with perimeter-type heating systems. One return may be adequate in smaller houses as shown in Fig. 84. In larger or split-level houses, return grilles are generally provided for each level or group of rooms. Locations of the returns within the space are not critical. They may be located in hallways, near entrance doors, in exposed corners, or on inside walls.

In the crawl space plenum system, the entire crawl space is used as an air-supply plenum or chamber. Warm air flows from near the ceiling to a central duct, is forced into the crawl space, and enters the rooms through perimeter outlets, usually placed beneath windows, or through continuous slots in the floor adjacent to the outside wall. With tight, well-insulated crawl space walls, this system can provide uniform temperatures throughout the house.

Houses built on concrete slabs may be heated by a perimeter-loop heating system (Fig. 86). Warm air is circulated by a counter-flow-type furnace through ducts cast in the outer edge of the concrete slab. The warm ducts heat the floor, and the warm air is discharged through floor registers to heat the room.

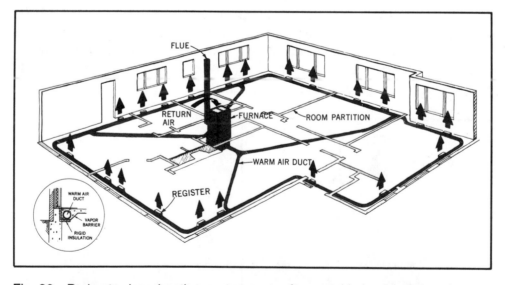

Fig. 86—Perimeter-loop heating systems are often used in basementless houses built on a concrete slab. The inset shows duct-slab-foundation construction details.

To prevent excessive heat loss, the edge of the slab should be insulated from the foundation walls and separated from the ground by a vapor barrier.

Warm-Air Heat Design

There are sound and substantial reasons why the warm-air furnace is the superior heating medium. The advantage over steam or hot water or radiant stoves is pure, fresh, warm air (when the fresh air inlet is taken from out-of-doors). If a warm-air furnace of the right size to properly warm a house is correctly installed, the air supplied will be healthful and pleasant.

With the room-by-room heat loss figures and the room-by-room heat gain figures, proceed as follows:

1. Select equipment by comparing Btu capacity of unit with requirements of the house (see Table 16).
2. Adjust Btu for each room if building requirements vary more than ten percent from equipment capacity.
3. Select register location, size, and number (see below) based on adjusted Btu for each room. Where system is to be used for both

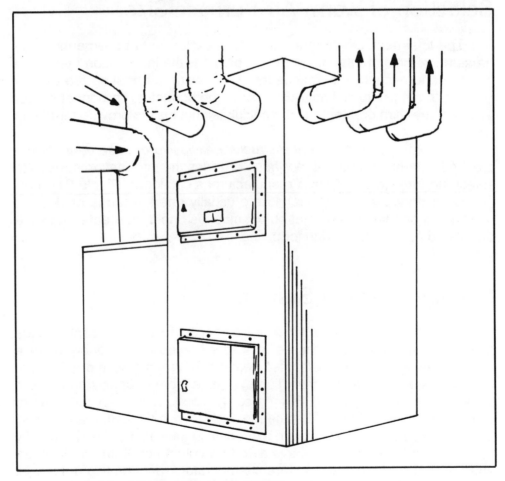

Fig. 87—Hand-fired Forked Warm Air Heater

heating and cooling, use the greater CFM equal to adjusted Btu for each room on heating and cooling (see Table 17).

4. Size branch pipe from chart based on step 3 CFM or Btu and locate trunk line (see Tables 18–20).

5. If trunk line is less than 30 feet, the use of extended plenum system is suggested.

 a. Add together CFM equal to Btu adjusted (heating or cooling, whichever greater) for all branch pipe coming off trunk line. Size trunk line from chart to carry total CFM.

6. For step-down trunk, start at end of trunk furthest from unit and add together CFM of branch pipes until step-up of size is desired. Select trunk size from chart.

7. For return air system repeat steps 3, 4, 5, and 6 above.

Selection of Warm-Air Furnace Size

The total heat loss from the structure, exclusive of basement, as expressed in terms of total Btu per hour, provides the information needed to select the furnace size. Furnaces for gravity-warm-air heating are usually rated in terms of Btu per hour (Btu/hr.) at the bonnet. (The bonnet output must be one-third greater than the calculated net loss to provide for piping loss.)

The *total heat loss from the structure exclusive of basement,* as expressed in terms of *total Btu per hour,* provides the information needed to select the furnace size from a manufacturer's catalogue (Table 21). Furnaces for gravity-warm-air heating are usually rated in terms of Btu per hour at the bonnet. The bonnet output must be one-third greater than the calculated net loss to provide for piping loss.

Gravity Warm-Air System

It is important to locate the coal stove or furnace in the center of the house and, preferably, in the cellar. Older homes will probably have a central chimney. A centrally located heater can offer a simple gravity convection system. A forced convection system is more expensive and more complex air flow is required. Forced convection is necessary with gas or oil but usually not with coal. Locate the coal stove in the cellar and initially cut some registers in the first floor. In order to induce a flow of air by gravity convection it is necessary to have a vertical column of air at one temperature surrounded by air at a different temperature, and the two masses of air have to be kept from mixing. Vertical ductwork should be used to keep

TABLE 16
DUCT AND ROUND PIPE SYSTEM DESIGN[1]

CFM	Heating Btu	Round Pipe	Square Duct	Floor Diffuser[2]	Sidewall Diffuser[3]	Baseboard Diffuser[2]	Return Grille[4]
32	2,400	4					
60	4,400	5			10×6	24″	
100	7,400	6	3¼ × 10	2¼ × 10		48″	10×6
120	8,900		3¼ × 12	2¼ × 12	12×6		12×6
145	10,700	7	3¼ × 14	4×14	14×6		14×6
180	13,300		8×6				
210	15,600	8					
270	20,000		8×8				24×6
290	21,500	9					
300	22,200						30×6
370	27,400		8×10				
390	28,900	10					
460	34,000		8×12				
500	37,000	11					
560	41,500		8×14				
620	45,900	12					
660	48,900		8×16				
770	57,000	13					
800	59,300		8×18				
900	66,700		8×20				
930	68,900	14					
1000	74,100		8×22				
1100	81,500		8×24				
1140	84,400	15					
1200	88,900		8×26				
1300	96,300	16	8×28				
1400	103,700		8×30				
1500	111,100	17	10×24				
1700	125,900		10×26				
1800	133,300	18					
1900	140,700		10×28				
2000	148,100		10×30				
2100	155,600	19					

[1]For systems up to 60 feet from unit to register. Based on .1 supply and .1 return and average number of fittings. For one or two long branch lines suggest one size larger branch pipe.

[2]Use proper size Floor Diffuser based on 6 to 7-½ Ft. throw based on actual CFM required.

[3]Use proper size Wall Diffuser based on Vertical throw of 6 to 7-½ ft. above floor based on actual CFM required.

[4]One full stud space must be provided for 10, 12 and 14″ grilles and two full stud spaces for 24 and 30″ grilles.

TABLE 17
RECOMMENDED VELOCITIES AND FREE AREAS OF RETURN AIR GRILLES ARE BASED ON CFM DELIVERY OF SUPPLY AIR TO ROOM

Warm air delivery to room, CFM	Air Velocity through Wall Grill				
	150 fpm	175 fpm	200 fpm	250 fpm	300 fpm
	Grille Free Area, sq. inches				
50	48	41	36	29	24
75	72	62	54	43	36
100	96	82	72	58	48
125	120	103	90	72	60
150	144	123	108	86	72
175	168	144	126	101	84
200	192	164	144	115	96
225	216	185	162	130	108
250	240	205	180	144	120
275	264	226	198	158	132
300	288	246	216	173	144
350	336	287	252	202	168
400	384	328	288	230	192

EXTENDED PLENUM SYSTEM

To determine size of main warm air duct, take number of branch runs (6″ pipe only) times 2 plus additional 2″ for duct width. Example: 7 branches × 2 = 14 plus 2 = 16″. Duct would be 16 × 8 for extended plenum system.

For an extended plenum containing angles add 1″ to the plenum for each 45° angle. Add 2″ for each 90° elbow used in the plenum. It is recommended that the extended plenum should not contain more than two angles or one elbow.

---

the warm and cold air from mixing with one another. The wider and taller that ductwork is, the better will be the resultant flow. In locating the furnace due consideration should be given to equalization of the length of leader pipes, spacing of the return air connections which should be such that the air will be uniformly distributed over the heat exchanger surface, minimization of the length of the smoke pipe, and ease of firing (in the case of solid-fuel-burning furnace). Face the furnace toward the fuel bin if practicable and provide adequate ventilation for combustion.

Floor outlets discharge air in a vertically upward direction. Perimeter outlets with a wide spread are recommended for heating. Warm air from the outlet blankets the cold surface of the room, traveling upward along

TABLE 18
GENERAL CHARACTERISTICS OF OUTLETS

Outlet Flow Pattern	Outlet Type	Preferred Location	Size Determined by
Vertical Spreading	Floor diffusers, baseboard, and low sidewall	Along exposed perimeter	Minimum supply velocity — differs with type and acceptable temperature differential
Vertical Non-spreading	Floor registers, baseboard, and low sidewall	Not critical	Maximum acceptable heating temperature differential
Low Horizontal	Baseboard and low sidewall	Long outlet at perimeter short outlet, not critical	Maximum supply velocity should be less than 300 fpm

TABLE 19
PREFERRED LOCATION OF RETURN INTAKES

Outlet Group	Heating Intake Location	Year-Round Intake Location
A—Spreading Vertical Jet	No Preference	Near Ceiling*
B—Non-spreading Vertical Jet	Near Floor	Near Floor
C—Horizontal Discharge, High	Near Floor	Near Floor
D—Horizontal Discharge, Low	No Preference	(System not recommended for year-round conditioning)

* Preferred location is near ceiling on opposite wall from outlet. However, the location is not critical.

TABLE 20
AREA AND CIRCUMFERENCE OF ROUND PIPES

Diameter of Pipe	Area in Square Inches	Circumference Round Pipe	Diameter of Pipe	Area in Square Inches	Circumference Round Pipe
3″	7.068	9.424	20″	314.16	62.83
4″	12.566	12.56	22″	380.13	69.11
5″	19.635	15.70	24″	452.39	75.39
6″	28.274	18.84	26″	530.93	81.68
7″	38.484	21.99	28″	615.75	87.96
8″	50.265	25.13	30″	706.86	94.24
9″	63.617	28.27	32″	804.24	100.5
10″	78.539	31.41	34″	907.92	106.80
12″	113.09	37.69	36″	1017.8	113.0
14″	153.93	43.98	38″	1134.1	119.3
16″	201.06	50.26	40″	1256.6	125.6
18″	254.46	56.54			

TABLE 21
TABLE FOR SELECTION OF FURNACE SIZE

Total Heat Loss in Btu per hour	Minimum Register Delivery in Btu per hour	Minimum Bonnet Output in Btu per hour
20,000	20,000	27,000
30,000	30,000	40,000
40,000	40,000	53,000
50,000	50,000	67,000
60,000	60,000	80,000
70,000	70,000	93,000
80,000	80,000	107,000
90,000	90,000	120,000
100,000	100,000	133,000
110,000	110,000	147,000
120,000	120,000	160,000
130,000	130,000	173,000
140,000	140,000	187,000
150,000	150,000	200,000

the exposed wall, across the ceiling, and down the far wall to form a relatively small stratified zone. Low sidewall outlets are commonly used to discharge warm air horizontally. During heating, the total air remains near the ceiling and, if the flow is sufficient, descends along a wall. The air flow rate must be relatively low to minimize drafts.

Room Air Distribution

If the air enters a room in such a way that either a draft or stratification results, the people in the room will not be comfortable. Neither draft nor stratification is apparent to the occupants of a comfortable room. The duct distribution system contains and directs the flow of air from the heating unit to the supply outlet. Final control of direction and velocity is provided by the supply outlet. After the air has passed through the outlet, the system can no longer affect the delivery pattern.

In moving warm air from the furnace, heated air rises rapidly but movement sideways (horizontally) is a problem. Distributing warm air to some part of the house becomes as important as moving cold air (return) to complete the heating system. It is essential to obtain a total circulation of warm and cold air through all areas of the house.

Even though a system delivers the required amount of warm air to a room, discomfort results if the air is not correctly distributed. A bonfire in the center of a room could supply enough Btu's to overcome the heat loss, but occupants would suffer from an excessively warm, toasted feeling on one side and cold radiation on the other.

How and where the air is admitted, the temperature differential between the supply air and the room air, temperature variations in the room, the velocity of the supply air, the number and characteristics of the supply outlets and return intakes, and the dimensions of the room are among the factors which must be successfully dovetailed in a warm-air heating system.

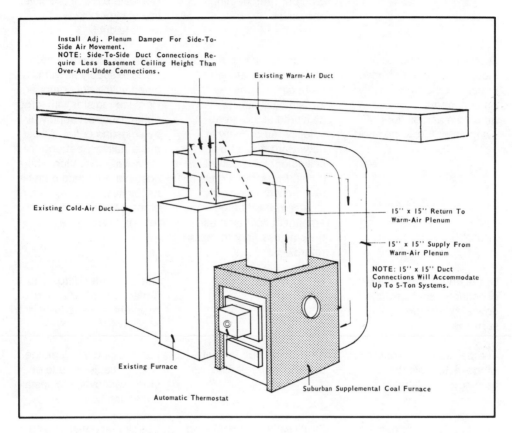

Fig. 88—Suburban's Coalmaster

Air Distribution Problems

Below is a table of air distribution problems (general and specific) in a warm air system, their causes and solutions. The listing is intended for use in solving localized problems, which may occur in a single room of a house.

TABLE 22

PROBLEM	CAUSE	SOLUTION
Insufficient heating, too little air flow through outlet or outlet air too low.	Duct damper closed; too small, incorrect outlet type or size; located in unconditioned space; too long, large air temperature drop between heating unit and outlet.	Adjust damper, rebalance system if necessary. Install larger or additional duct. Check outlet. Replace if necessary. Insulate duct. Relocate outlet if possible. Install larger duct or insulate duct
Drafts in occupied areas, incorrect spread or throw, supply air blowing directly on occupants, or excessive air drawn across floor of living area, throw too short.	Characteristics of outlet are not correctly matched to room dimensions. Air from low sidewall outlets delivered at too great velocity. Air delivery at downward angle from ceiling or high sidewall outlet. Projected air hits obstruction. Incorrectly located return. Large amount of cold air must pass across floor to reach return.	Adjust or replace outlet. Reduce supply air velocity by adjusting duct damper. Replace register with larger outlet. Redirect supply air by adjusting outlet. Adjust outlet delivery pattern. Remove obstruction where possible. Relocate outlet if necessary. Relocate return or increase number of returns in structure.
Drafts under large glass exposures, cool air drifts down from window surfaces.	Incorrect outlet.	Use perimeter diffuser to project air upwards and counteract descending drift of cool air.
Temperature near doors or windows too low (for heating); air infiltration	Air leakage due to faulty building construction.	Locate source of leakage. Seal by caulking, use of storm windows, etc. Install outlet under window.
Occupants feel "closed-in"—air is stale; large temperature variations in occupied spaces	Stratification boundary too high. (Stratified zone for heating lies between stratification boundary and floor.)	Check that outlet spread, throw, angle of delivery, etc., are correct for intended air pattern. Increase air flow through outlet.
Room air too warm; widely varying room air temperatures	Incorrect placement of thermostat; responds to sun entering windows, change in wind direction and/or velocity on exposed areas of structure, inside loading effects (cooking, washing, bathing), etc.	Change location of thermostat. Check need for zoning.

Cold floor; insufficient heat supplied at floor level	Duct damper is closed or system does not provide for floor warming.	Distribution system should provide floor warming. Make sure duct damper is open and sufficient heat is being supplied at or below floor level.

Location of Registers

A gravity-warm-air heating system consists of more than furnace alone, registers alone, or a set of controls. A properly installed and operated system is a closely knit working combination of all component parts. Some simple basic considerations should be made for proper installation of registers. The free passage of the warm air discharged by the coal furnace is of vital importance. The essential thing is air capacity; that is, an eight-by-ten register should be considered as having 40 square inches net air capacity. In a gravity-warm-air system it is absolutely necessary to use larger sized registers than are commonly used for forced air systems. The table given below is a safe guide to determine the correct size of registers to use with the standard sizes of pipe listed for a gravity-warm-air system. When cold air is circulated back from room to furnace, face plates of greater capacity than the return pipe should be used. This is necessary because the suction of the furnace does not draw the air above the register with same force it exerts when air has entered the return pipe, and a large air opening is therefore required.

TABLE 23
TABLE OF AREAS

Dimension of Pipe	Area in Square Inches	Size of Register Required
8″	50	8 × 12
9″	63	9 × 14
10″	78	10 × 16
12″	113	14 × 16
14″	154	16 × 20
16″	201	18 × 24

Warm-Air Registers and Pipes

Warm-air registers are usually installed in the baseboard, in the floor, or (if stack area is ample) in the sidewall just above the baseboard. Where registers are installed in the baseboard, the throats of register boxes and boots must be equal in area to that of the connecting pipe. Floor registers are easy to install, but they may interfere with placement of carpets, and may be dirt-catchers.

Warm-air risers must be located in inside warm walls and not in outside or cold partition walls. Stacks, or risers, to second-story rooms should be amply sized, since the stack is usually the most restricted section of a warm-air room.

Warm-air leader connections should be uniformly distributed around the furnace bonnet (avoiding connections over the firing door in the case of coal-fired furnaces). The top of all the leader pipes should be as near the top of the bonnet as possible, and the leader pipes should be at the same level.

Warm-air-supply pipes usually consist of round leaders pitched upward from furnace bonnet to the boot connections; round leaders, all rising abruptly at furnace bonnet and running level to boot connections; gravity casing with plenum chamber and round leaders running level to boot connections (with required clearances); or gravity casing with plenum chamber and extended plenum type of trunk, complete with short leaders (branches) to boot connections.

Cold-Air-Return Intakes

Two or more cold-air-return intake ducts are commonly used. In the case of a small compact house with a central hall, a single return may be used, but not in a closed vestibule.

Returns should be short, direct, and with the least number of turns and without restrictions. A lack of restriction of return capacity can be a major source of trouble.

Cold-air-return intakes may be placed at either warm wall or cold wall locations. If ducts can be made reasonably short and direct, cold wall locations are usually preferred. Cold-air-return intakes are preferably placed in the floor. If baseboard returns are used, throat area equal to the area of the return air intake must be provided. The top of the openwork of the return air intake must not extend more than 14 inches above the floor and the bottom must be within one inch of the floor. Return ducts from second-story rooms are not as effective relatively as short returns from the first story.

The return-air duct should not be heated along any part of its run. Return-air shoe connections should enter the casing below the grate of fire level. On all return pipes larger than 12 inches a transition shoe should be used.

Selected Warm-Air Heating Systems

Most warm-air furnaces (hand-fired and automatic) using coal work on the same basic principles, using heat from the coal fire to warm air in an overhead metal box called a plenum. The heated air is then conducted from the plenum to various locations in the house through a system of galvanized steel ducts. Most box coal stoves can be used as warm air furnaces but will require a sheet-metal box and/or plenum configuration. Some warm-air furnaces have blowers that enhance the air flow, while others use the gravity method or are augmented by a separate forced air system. The following warm air stoves and furnaces represent only a few of the models available:

Riteway's[1] Models 37 and 2000 are hot air furnaces and can produce 50,000 to 73,000 Btu of heat per hour, enough for an average-size, insulated house. Riteway has larger models (LF–20–30–50–70) which include draft inducer, ash pit blower, heat blower, barometric draft regulator, fan limit control, water heater adaptors, water heaters, and thermostat. Fig. 89 illustrates Riteway coal furnace configuration.

The basic design of Hunter Thermomax furnaces[2] (22 inches–140,000 Btu; 24 inches–175,000 Btu; and 27 inches–225,000 Btu) was actually developed many years ago, when solid fuels such as coal were the rule rather than the exception. The popularity of the remarkably rugged construction is proven by the fact that more than 1,000,000 large-capacity domestic furnaces of this design were produced and sold, according to the company.

Anticipating the return to coal as increasingly popular fuel, the Hunter Energy Company obtained the original dies required to produce these heavy-duty furnaces for today's homes.

For example, the combustion chamber and drum head of the Thermomax furnace are blanked from single pieces of heavy seven-gauge steel on a giant press. The combustion chamber is blanked, riveted, and welded for strength and tightness.

The unique clam shell radiator—claimed to be the key to the Thermomax's remarkable heat-producing efficiency—is constructed from two pieces of 12-gauge steel, drawn and welded.

Recognizing that the basic reason for heating with coal is economy, the Thermomax provides a large capacity and high efficiency, with up to ten hours between loadings of wood and up to sixteen hours when used as a coal furnace.

Thermomax furnace features are:

1. *Drum head* and pouch formed in one piece from heavy seven-gauge steel.

---

[1]Riteway Manufacturing Company, Harrisonburg, VA 22801
[2]Hunter Energy Inc., 4900 Southway SW, Canton, OH 44706

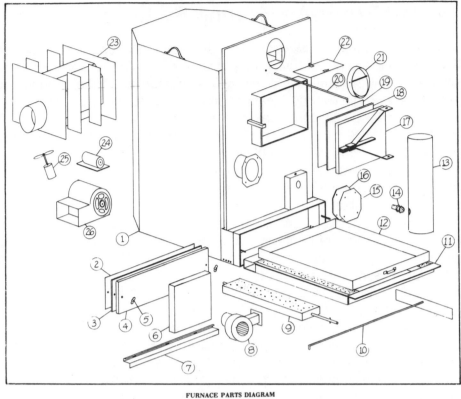

FURNACE PARTS DIAGRAM

1. Body assembly	10. Fuel selector damper	19. Fuel door liner
2. Ash door liner	11. Grate frame	20. Operating rod
3. Ash door gasket	12. Ash pan	21. Barometric draft regulator
4. Ash door	13. Combustion flue	22. Direct draft damper
5. Wing nuts	14. Secondary air distributor	23. Heat exchanger
6. Firebrick	15. Mounting tube cover	24. Heat blower motor
7. Brick retainer	16. Mounting tube gasket	25. Draft inducer
8. Ash pit blower	17. Fuel door and latch assembly	26. Heat blower
9. Grate bar	18. Fuel door gasket	

Fig. 89—Furnace parts diagram

2. *V-shaped radiator baffle* which deflects the hot gases so that all parts of the radiator become effective heating surfaces.
3. *Foil-faced fibreglass insulation* that reduces heat loss, improves efficiency, and cools the outer casing.
4. *Combustion chamber* blanked and fabricated from one piece of nine-gauge steel, riveted and welded for strength and tightness.
5. *Fire pot* brick, which provides large fuel capacity.
6. *Duplex roller-bearing grates* made of heavy-duty cast iron with rotating outer ring and center dump bars.
7. *Ash pit* with 12-gauge steel base formed in one piece with base securely welded to drum. Ash pit is dust- and gas-tight.
8. *Large-capacity die-formed radiator* made from two pieces of securely welded 12-gauge steel.

9. *Handle for rotating grates* which permits ash removal without opening fire door.

Combustion gases pass from the combustion chamber through the baffled radiator and out of the flue pipe. Cold air enters at the side of the furnace and rises naturally or is forced by the blower, past the radiator and combustion chamber to enter the warm-air duct system.

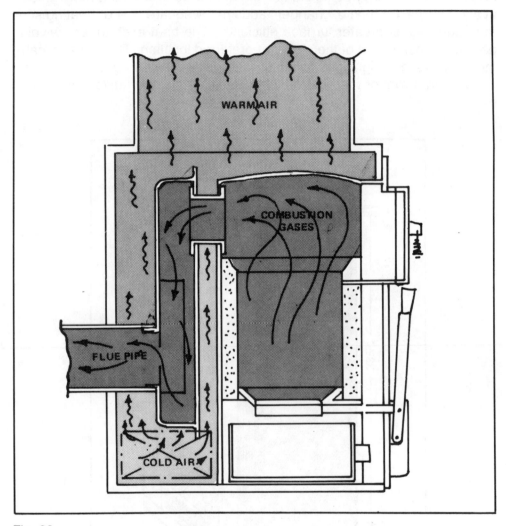

Fig. 90

The Combustioneer (Model 24FA with stoker S20) manufactured by Will-Burt[3] has a stoker firing rate of 20 pounds per hour and the Btu at bonnet (based on 12,000 Btu coal) is 156,000. The warm air plenum is 37 × 32-¾ inches in size; a one-quarter-horsepower motor blower.

[3] The Will-Burt Company, 169 S. Main Street, Orrville, OH 44667

The firebrick lining is made to withstand heat up to 3050°F., says the manufacturer. There are no expensive castings to crack or burn out. This design makes an ideal firing chamber. The single-piece, heavy steel base provides a rigid foundation for the furnace. The furnace body is made of heavy boiler-plate steel and is precision welded, absolutely smoke- and gas-tight. The large, boiler-plate steel, doughnut-type radiator is welded in one piece. The heavy baffle causes the hot gases to circulate more effectively. This "extra heat exchanger" adds many square feet of heating surface and means greater furnace efficiency. The heat exchanger swivels, permitting installation of the stoker at preferred location. The furnace cabinet panels are designed so no modification is needed for right- or left-hand installation. Front or rear installation will require slight modification of one panel.

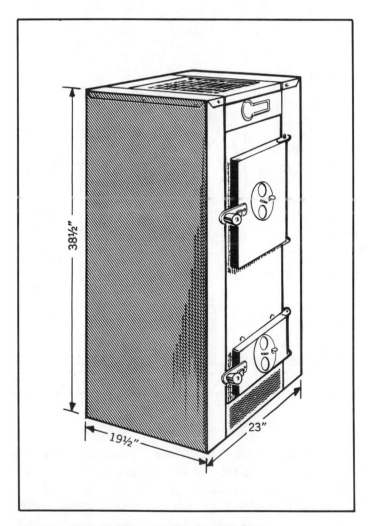

Fig. 91—35R Radiant Heater by Will-Burt

Other Will-Burt models, such as the 65 R/M, have a blower to increase the flow of warm air.

Keystoker[4] forced-warm-air heating units can provide automatic heating at the lowest possible cost. Use anthracite coal and eliminate the time and work demanded by hand-feeding. A large, silent air-circulating blower will assure complete circulation of warm, filter-cleaned air throughout every room in the house. The stoker-furnace specifications include: three controls (room thermostat, timer, combination bonnet control); blower and motor; fiberglass filters; and one ash receptable.

TABLE 24

Model No.	Bonnet Output Btu	Length	Height	Width
A80	80,000	63¼	41¼	27
A125	125,000	78½	51¼	30¾
A200	200,000	86⅝	56⅜	36¾
A300	300,000	69¼	67	40

The Keystoker warm-air furnace is built to deliver maximum heat at the lowest initial investment and lowest operational cost. The downdraft heat exchanger increases heating surface, permitting direct circulation of every thermal unit of heat into the warm-air leader pipes for rapid distribution. The heating surface is properly proportioned to the grate area, resulting in appropriate balance and greater heat delivery per pound of fuel burned.

The blower has been expertly engineered and designed to give many trouble-free years of silent operation. Sealed pit ash removal assures clean, dust-free operation.

The Superheater Mark II[5] space heater, an automatic coal-fired warm-air heater with 1380 CFM air circulation, output Btu (75,000 Btu/hr.) has these standard features:

Packaged design: Heat exchanger, stoker-hopper, and control system are uniquely combined into a single appliance for easy installation and operation. Handsome cabinet with baked enamel finish will complement any decor.

Finned heat exchanger: Engineered for maximum heat recovery and formed of heat-resistant high-alloy steel. Outside fins dissipate heat over entire heat exchanger, virtually eliminating hot spots and extend service

---

[4]Keystone Manufacturing Company, Schuylkill Haven, PA 17972
[5]Stokermatic Corporation, 1610 Industrial Road, Salt Lake City, UT 84104

life. Fins add to strength and reduce weight and size of heat exchanger with an increase of efficiency.

Drive unit: The drive unit has been time-tested to provide quiet, efficient, and reliable service. Motor, gear box, and combustion fan are integrated into one unit to save space and increase efficiency.

Under-feed burner: An under-feed burner combined with an over-fire air system enables a unique balancing of fuel and air for complete and efficient combustion.

Simple, economical installation: Unit is completely prewired and tested at the factory. The heater can be vented into a suitable chimney. It is insulated and compact so it can locate close to a wall, thus taking a minimum of floor space.

Easy-out clinker bucket: Built-in receiver for clinkers and ashes rests in front of heat exchanger, lifts out for clean, easy handling without spills.

Sealed coal hopper: Hopper is coated with a protective sealant to suppress noise and resist corrosion. Gasket on hopper lid prevents dust or smoke emissions. Holds 225 pounds of coal, enough for almost 24 hours of continuous firing with average Btu coal.

Floating feed screw: Engineered to permit rocks and most debris to pass through the feed tube without stalling. The feed screw is shear-pin-equipped for added protection and can be withdrawn to remove an obstruction.

All-season air circulation: Fan can be manually activated for quiet air circulation during warm summer months.

Over-fire air jet: Directs air into the center of flame forcing it towards fast-absorbing primary heating surfaces. Uniquely balances fuel and combustion air for complete and efficient combustion.

Safety and economy: The Superheater Mark II can provide more heat per dollar with virtually no smoke or wasted fuel, the company claims. The insulated cabinet is safe to touch and the circulating fan is enclosed to keep children's fingers from being hurt.

All Stokermatic models are furnished with Underwriters' Laboratories-listed controls and electrical components. Controls are standardized among models to utilize the highest quality components available.

Glossary

Agglomerating Index: A measurement that roughly indicates the caking properties of coal.

Air Cleaner: A device designed to remove air-borne impurities such as dusts, fumes and smokes. (Air cleaners include air washers and air filters.)

Air Conditioning: The simultaneous control of all or at least the first three of those factors affecting both the physical and chemical conditions of the atmosphere within any structure. These factors include temperature, humidity, motion, distribution, dust, bacteria, odors, toxic gases, and ionization, most of which affect in greater or lesser degree human health or comfort.

Air Infiltration: The in-leakage of air through cracks and crevices, and through doors, windows and other openings, caused by wind pressure or temperature difference.

Air Washer: An enclosure in which air is forced through a spray of water in order to cleanse, humidify, or dehumidify the air.

Ambient Air Quality Standards: According to the Clean Air Act of 1970, the air quality level which must be met to protect the public health (primary) and welfare (secondary). Secondary standards are more stringent than Primary Ambient Air Quality Standards.

Anthracite: A hard, compact variety of natural coal, of high luster, differing from bituminous coal in that it contains only a small amount of volatile matter, in consequence of which it burns with a nearly non-luminous flame.

Aquifer: A subsurface zone that yields economically important amounts of water to wells; a water-bearing stratum or permeable rock, sand, or gravel.

Ash: The non-combustible residue left after coal is burned completely, e.g., silica, alumina, magnesia, lime, iron oxide, sodium, and potassium.

Ashes: The term generally applied to the residue obtained from the combustion of coal, and which may contain some combustible matter (see ash).

Ash-Free Basis: In order to reduce the number of variables in comparing one coal with another, it is standard procedure to calculate the composition of a coal on an ash-free basis. For example, a coal with a ten percent ash, 5 percent volatile matter, and eighty-five percent fixed carbon, would contain

$$\frac{5}{90} \times 100\% = 5.55 \text{ percent volatile matter, and}$$
$$94.45 \text{ percent fixed carbon,}$$
on an ash-free basis.

Ash Fusing Point: The temperature at which the ash begins to soften, as determined by a procedure adopted by the A.S.T.M. Ashes which fuse in the range 1900° to 2200° F. are considered low fusing; those in the range 2200° to 2600° F., medium fusing; and those from 2600° to 3100° F. or over, are considered non-fusing.

Atmospheric Pressure: This is the pressure due to the weight of the atmosphere. Normal atmospheric pressure is 14.7 pounds per square inch at sea level. The average atmospheric pressure for any locality depends on the altitude above sea-level.

Base Load: The minimum load of a utility, electric or gas, over a given period of time.

Bituminous Coal: The coal ranked below anthracite. It generally has a high heat content and is soft enough to be readily ground for easy combustion. It accounts for the bulk of all coal mined in this country.

Black Lung Disease: A group of pulmonary diseases that are common among coal miners.

Bone: Slaty coal, or rock with a high carbon content ranging from 40 percent to 75 percent fixed carbon. It is neither coal nor rock but has some characteristics of each.

British Thermal Unit (Btu): Generally used in engineering work in the United States and Great Britain, a Btu is 1/180 of the heat required to raise the temperature of one pound of water from 32° F. to 212° F. It formerly was defined as the heat required to raise one pound of water from 62° to 63°F. One Btu will raise 55 cubic feet of air 1° F.

Calorific Value (Heat Value): The heat in Btu's generated by the complete combustion of one pound of a solid or liquid fuel of one cubic foot of gaseous fuel at standard pressure and temperature.

Carbon: The principal constituent of coal and most other fuels, whether in the solid, liquid or gaseous form.

Clinker Formation: The tendency toward clinker formation is roughly proportional to the fusibility of the ash.

Clinkers: A mass of fused or molten ashes.

CO: Symbol for carbon monoxide, a product resulting from incomplete combustion of the carbon in the fuel.

CO_2: Symbol for carbon dioxide, the product resulting from complete combustion of the carbon in the fuel.

Coal Caking: Coal which fuses and becomes a pasty mass when heated.

Coal Free Burning (Non-Caking): Coal which does not fuse together and become a pasty mass when heated, but burns freely.

Coal Gasification: The process that produces synthetic gas from coal.

Coal Liquefaction: Conversion of coal to a liquid for use as synthetic petroleum.

Coefficient of Transmission: The amount of heat (Btu) transmitted from air to air in one hour per square foot of the wall, floor, roof or ceiling for a difference in temperature of 1° F. between the air on the inside and that on the outside of the wall, floor, roof or ceiling.

Collier: A coal miner or a ship for carrying coal.

Colliery: A coal mine and its buildings, equipment, etc.

Color (and Luster): Coals classified as bituminous are considered black in color. There are in these coals bands and laminations of varying shades and luster. Coal in the broad sense of the word covers a wide range of colors. The color range begins with lignites that can have varying shades of brown to brilliant black. Bituminous and anthracite coals range from black-gray to the pure black that sometimes has blue and yellowish tints.

Combustible Substance: One which, when heated to its temperature of ignition, readily unites with oxygen, resulting in the production of heat.

Combustion: The chemical union of a combustible substance with oxygen, resulting in the production of heat.

Combustion Chamber: That part of the boiler in which gases from the coal unite with additional oxygen to further combustion.

Combustion, Rate of: The number of pounds of coal burned per square foot of grate per hour.

Condensation: The change of a substance from a vapor into a liquid state due to cooling.

Concealed Radiator: See "Convector."

Conductance: The amount of heat (Btu) transmitted from surface to surface in one hour through one square foot of a material or construction, whatever its thickness, when the temperature difference is 1° F. between the two surfaces.

Conduction: The transmission of heat through and by means of matter unaccompanied by any obvious motion of the matter.

Conductivity: The amount of heat (Btu) transmitted in one hour through one square foot of a homogeneous material one inch thick for a difference in temperature of 1° F. between the two surfaces of the material.

Control: Any manual or automatic device for the regulation of a machine to keep it at normal operation. If automatic, it is considered that the device is motivated by variations in temperature, pressure, time, light, or other influences.

Convection: The transmission of heat by the circulation of a liquid or a gas such as air. Convection may be natural or forced.

Convector: A concealed radiator. A heating unit and an enclosure or shield located either within, adjacent to, or exterior to the room or space to be heated, but transferring heat to the room or space mainly by the process of convection. If the heating unit is located exterior to the room or space to be heated, the heat is transferred through one or more ducts or pipes.

Degree-Day: A unit, based upon temperature difference and time, used in specifying the nominal heating load in winter. For any one day there exist as many degree-days as there are Fahrenheit degrees' difference in temperature between the average outside air temperature, taken over a 24-hour period, and a temperature of 65° F.

Dehumidify: To remove water vapor from the atmosphere; to remove water vapor or moisture from any material.

Density: The weight of a unit volume, expressed in pounds per cubic foot:

$$d = \frac{W}{V}$$

Dew-Point Temperature: The temperature corresponding to saturation (100 percent relative humidity) for a given moisture content.

Diffuser: A vaned device placed at an air supply opening to direct the air flow.

Dry Basis: Samples of coal "as received" vary considerably in moisture content. To eliminate this variable, analyses usually are calculated on a "dry basis," reporting the amounts of the constituents as a given percentage of dry coal, moisture free.

Elasticity: The fractional change in a variable that is caused by a unit change in a second variable. Income elasticities are important in energy estimates, since these estimate the changes in quantities of energy demanded as incomes change.

Fines: Coal two sizes smaller than the size specified as shipped or purchased.

Fixed Carbon: The combustible portion of dry coal which is left after the volatile matter has been driven off in the standard analytical procedure for proximate analysis.

Flue-Gas Desulfurization: The use of a stack scrubber to reduce emissions of sulfur oxides.

Fluidized Bed: A fluidized bed results when gas is blown upward through finely crushed particles. The gas separates the particles so that the moisture behaves like a turbulent liquid. Being developed for coal burning for greater efficiency and environmental control.

Fly ash: Light-weight solid particles that are carried into the atmosphere by stack gases.

Fracture: Fracture is the characteristic way in which coal breaks, displaying its texture. The nature of the fracture in coal is often useful in distinguishing one variety of coal from another. Anthracite breaks with a conchoidal fracture, i.e., curving surface; bituminous and semi-bituminous coals break with a blocky or cubical fracture; and lignites and sub-bituminous coals are variable in fracture and in the resultant shape of lump.

Free Swelling Index (F.S.I.): The free swelling test is one in which a one-gram sample of coal is heated in a silica crucible over a gas burner at a prescribed rate to form a coke button. It is used as an indication of the coking characteristics of the coal when burned as a fuel.

Friability: The friability of a coal is its tendency to degrade in size when subjected to handling and is of special importance in determining the relative size consist of various coals when delivered to the fuel bed in various types of burning equipment. It has no relation to grindability.

Fritted ash: Soft, formed ash.

Grate Area: The area of the grate surface, measured in square feet, to be used in estimating the rate of burning fuel.

Greenhouse Effect: The potential rise in global atmospheric temperatures due to an increasing concentration of CO_2 in the atmosphere. CO_2 absorbs some of the heat radiation given off by the Earth, some of which is then reradiated back to the Earth.

Grindability: Ease of pulverizing of coals, determined in comparison with a coal chosen as 100 grindability.

Gross Energy Demand: The total amount of energy consumed by direct burning and indirect burning by utilities to generate electricity.

Humidify: To add water vapor to the atmosphere; to add water vapor or moisture to any material.

Humidity: The water vapor mixed with dry air in the atmosphere. Absolute humidity refers to the weight of water per unit volume of space occupied, expressed in grains or pounds per cubic foot. Specific humidity refers to the weight of water vapor in pounds carried by one pound of dry air. Relative humidity is a ratio, usually expressed in percent, used to indicate the degree of saturation existing in any given space resulting from water vapor present in that space.

Ignition Temperature: The temperature to which a combustible must be raised to cause a rapid chemical union with oxygen.

In Situ Processing: In-place processing of fuel by combustion without mining; applies to oil shale and coal.

Joule: A unit of energy which is equivalent to one watt for one second. One Btu = 1,055 Joules.

Lignite: The lowest rank coal from a heat content and fixed carbon standpoint.

Metallurgical Coal: Coal used in the steelmaking process. Its special properties and difficulty of extraction made it more expensive than steam coal.

Methane: CH_4, carburated hydrogen or marsh gas formed by the decomposition of organic matter. It is the most common gas found in coal mines.

Mineral Matter Free: A term applied to pure coal substance. It is calculated from a prescribed formula and is used as a basis of comparison for various coals.

Moisture: Essentially water, quantitatively determined by drying the coal at 105° C. for one hour.

Moisture Free: Since the moisture possesses no heating value, it is customary, by calculation, to reduce weight of coal, analysis, etc. to the values they would have if the coal was completely dry.

Oil Shale: A finely grained sedimentary rock that contains an organic material, kerogen, which can be extracted and converted to the equivalent of petroleum.

Particulate Matter: Solid airborne particles, such as ash.

Peak Power: The maximum amount of electrical energy consumed in any consecutive number of minutes, say 15 or 30 minutes, during a month.

Plenum Chamber: An air compartment maintained under pressure and connected to one or more distributing ducts.

Porosity: Porosity is the degree to which coal is permeable by liquids. All coal from bituminous down to lignite is porous, and the degree of porosity is greater in the lower rank coals. Liquids are absorbed by coal in two ways: through the minute capillary cells in the coal structure and through the small cracks and fissures of the cleats and bedding planes of the seam formation.

Proximate Analysis of Coal: The proximate analysis is a set of numerical values obtained when a coal is subjected to a series of purely arbitrary procedures which have been standardized by the A.S.T.M. The moisture is determined by drying the coal at 105° C. for one hour. The ash is obtained as a residue on burning the coal. The loss of weight on heating the coal seven minutes at 950° C., minus the moisture, is called the volatile combustible matter; the carbon residue, minus the ash, is called the fixed carbon. The total sulfur is usually determined in connection with the proximate analysis.

Psychrometer: An instrument for ascertaining the humidity or hygrometric state of the atmosphere.

Radiation: The transfer of energy from point to point in space by means of waves. For example, energy from the sun warms the earth by radiation.

Radiator: A heating unit exposed to view within the room or space to be heated. A radiator transfers heat by radiation to objects it can "see" and by conduction to the surrounding air which in turn is circulated by natural convection; a so-called radiator is also a convector but the single term "radiator" has been established by long usage.

Reserves: Resources of known location, quantity, and quality which are economically recoverable using currently available technologies.

Retrofit: A modification of an existing structure, such as a house or its equipment, to reduce energy requirements for heating or cooling. There are basic types of retrofit: equipment, such as a heat pump, replacing less efficient equipment; and insulation, storm doors, caulking, etc., designed to lower energy requirements.

Saturated Air: Air containing as much water vapor as it can hold without any condensing out; in saturated air, the partial pressure of the water vapor is equal to the vapor pressure of water at the existing temperature.

Seam: A bed of coal or other valuable mineral of any thickness.

Size Stability: Size stability of coal is its tendency to resist degradation in size when subjected to handling at the mine, during transit to the consumer and in storage.

Slate: A dense, fine-grained rock produced by the compression of clays, shales, and certain other rocks, so as to develop a characteristic cleavage, which may be at any angle with the original bedding plane; also defined as any material which has less than 40 percent fixed carbon.

Specific Gravity: The ratio of the weight of a body to the weight of an equal volume of water at some standard temperature, usually 39.2 F.

Stack Scrubber: An air pollution control device that usually uses a liquid spray to remove pollutants, such as sulfur dioxide or particulates, from a gas stream by absorption or chemical reaction. Scrubbers are also used to reduce the temperature of the emissions.

Steam Coal: Coal suitable for combustion in boilers. It is generally soft enough for easy grinding and less expensive than metallurgical coal or anthracite.

Sub-bituminous Coal: A low-rank coal with low fixed carbon and high percentages of volatile matter and moisture.

Sulfates: A class of secondary pollutants that includes acid-sulfates and neutral metallic sulfates.

Sulfur: An element that appears in many fossil fuels. In combustion of the fuel the sulfur combines with oxygen to form sulfur dioxide.

Sulfur Dioxide: One of several forms of sulfur in the air; an air pollutant generated principally from combustion of fuels that contain sulfur.

Synthetic Fuel: A fuel produced biologically, chemically, or thermally transforming other fuels or materials.

Ultimate Analysis: The analysis which gives the relative amounts of the elements present in the coal. Elements usually included are carbon, hydrogen, oxygen, nitrogen, sulfur, and ash. The composition of the ash is usually not determined.

Volatile Matter: Volatile matter is the portion of the coal substance which may be driven off by heat as a combustible gas before the coal starts to burn. It is determined analytically and is reported in the proximate analysis.

Western Coal: Can refer to all coal reserves west of the Mississippi. By Bureau of Mines definition, it includes only those coalfields west of a straight line dissecting Minnesota and running to the western tip of Texas. Wyoming and Montana (sub-bituminous) and North Dakota (lignite) have the largest reserves.

Appendix I
Coal Information Sources

Federal Agencies

Bartlesville Energy Research Center
PO Box 1398
Bartlesville, OK 74003

Grand Forks Energy Research Center
PO Box 8213
Grand Forks, ND 58201

Morgantown Energy Research Center
PO Box 880
Morgantown, WV 26505

Pittsburgh Energy Research Center
4800 Forbes Avenue
Pittsburgh, PA 15213

US Geological Survey
National Center
Reston, VA 22092

Selected State Agencies[1]

Geological Survey of Alabama
PO Drawer O
University, AL 35486

Arkansas Geological Commission
3815 W. Roosevelt Road
Little Rock, AR 72204

Arkansas Department of Labor
Mine Inspection Division
250 Central Mall
Fort Smith, AR 72901

Colorado Geological Survey
1313 Sherman Street, Room 715
Denver, CO 80203

Illinois State Geological Survey
121 Natural Resources Building
Urbana, IL 61801

Illinois Department of Mines and Minerals
704 State Office Building
400 S. Spring Street
Springfield, IL 62706

Indiana Geological Survey
611 N. Walnut Grove
Bloomington, IN 47401

Indiana Board of Mines and Mining
1119 Wabash Avenue
Terre Haute, IN 47807

Iowa Geological Survey
123 N. Capitol Street
Iowa City, IO 52242

Kansas Geological Survey
1930 Avenue A, Campus West
University of Kansas
Lawrence, KS 66044

[1]Most State Energy Units or Offices have coal resource information.

Kentucky Geological Survey
307 Mineral Industries Building
University of Kentucky
Lexington, KY 40506

Kentucky Department of Mines and Minerals
PO Box 680
Lexington, KY 40501

Maryland Geological Survey
Merryman Hall—Johns Hopkins University
Baltimore, MD 21218

Maryland Bureau of Mines
City Building
Westernport, MD 21562

Missouri Geological Survey
PO Box 250
Rolla, MO 65401

Missouri Department of Labor & Industrial
    Relations
Inspection Section
722 Jefferson Street
Jefferson City, MO 65101

Montana Bureau of Mines and Geology
Montana College of Mineral Science &
    Technology
Butte, MT 59701

New Mexico State Bureau of Mines & Mineral
    Resources
New Mexico Tech
Socorro, NM 87801

New Mexico State Mine Inspector
National Building
505 Marquette Avenue, NW
Albuquerque, NM 87101

New Mexico Coal Surface Mining Commission
State Capitol
Santa Fe, NM 87501

North Dakota Geological Survey
University Station
Grand Forks, ND 58201

North Dakota Department of Safety
Workmen's Compensation Bureau
Russel Building, Highway 83 North
Bismarck, ND 58505

Ohio Geological Survey
Fountain Square, Building 6
Columbus, OH 43224

Ohio Division of Mines
2323 W. Fifth Avenue
Columbus, OH 43216

Oklahoma Geological Survey
University of Oklahoma
830 Van Vleet Oval, Room 163
Norman, OK 73069

Oklahoma Department of Mines
117 State Capitol
Oklahoma City, OK 73105

Pennsylvania Department of Environmental
    Resources
PO Box 1467
Harrisburg, PA 17120

Tennessee Division of Geology
G-5 State Office Building
Nashville, TN 37219

Tennessee Department of Labor
Division of Mine Inspection
306 State Office Building Annex
618 Church Street, SW
Knoxville, TN 37902

Utah Geological and Mineral Survey
606 Black Hawk Way
Salt Lake City, UT 84108

Division of Mineral Resources
PO Box 3667
Charlottesville, VA 22903

Virginia Department of Labor and Industry
Division of Mines and Quarries
Big Stone Gap, VA 24219

Washington Department of Natural Resources
Geology and Earth Resources Division
Olympia, WA 98504

West Virginia Geological & Economic Survey
PO Box 879
Morgantown, WV 26505

West Virginia Department of Mines
State Capitol, Room E-151
Charleston, WV 25305

Geological Survey of Wyoming
PO Box 3008, University Station
Laramie, WY 82071

Wyoming Mine Inspector's Office
PO Box 1064
Rock Springs, WY 82901

Independent Coal Organizations

Alabama Mining Institute
1703 John A. Hand Building
Birmingham, AL 35203

Alabama Surface Mining & Reclamation Council
244 Goodwin Crest, Suite 318
Executive Office Park
Birmingham, AL 35209

Alabama Surface Mining Environmental
    Association
Bickerstaff Clay Products Co., Inc.
PO Box 1178
Columbus, GA 31907

American Coal Association
343 S. Dearborn Street, Suite 608
Chicago, IL 60605

Anthracite Institute
425 Candlewyck Road
Camp Hill, PA 17011

Arkansas-Oklahoma Coal Operators
    Association
PO Box 186
Fort Smith, AR 72901

Association of Bituminous Contractors
2020 K Street, NW, Suite 800
Washington, DC 20006

Association Technique de L'importation
    Charbonniere
Delegation in North America
1 World Trade Center, Suite 5201
New York, NY 10048

Big Sandy-Elkhorn Coal Operators Association
PO Box 7223
Lexington, KY 40502

Bituminous Coal Operators' Association
918 16th Street, NW
Washington, DC 20006

Bituminous Coal Research, Inc. (BCR)
350 Hochberg Road
Monroeville, PA 15146

Carnegie-Mellon University
Coal Research Laboratory
Pittsburgh, PA 15213

Central Pennsylvania Coal Producers'
    Association
Masonic Building, Suite 219
Ebensburg, PA 15931

Citizens' League to Protect the Surface Rights
Blackey, KY 41804

City College of New York
City College Clean Fuels Institute
Department of Chemical Engineering
New York, NY 10031

Coal Exporters Association of the United States
1130 17th Street, NW
Washington, DC 20036

Coal Mining Institute of America
416 Ash Street
California, PA 15419

Coal Operators and Associates, Inc.
315 Hopkins Building
Pikeville, KY 41501

Coal Producers Association of Illinois
314 E. Adams Street
Springfield, IL 62701

Colorado & New Mexico Coal Operators
   Association
PO Box 1394
Englewood, CO 80110

Eastern Bituminous Coal Association
Masonic Building, Suite 219
Ebensburg, PA 15931

Greenbrier Coal Operators Association
PO Box 185
Quinwood, WV 25981

Harlan County Coal Operators Association
PO Box 230
Harlan, KY 40831

Hazard Coal Operators Association
Citizens Bank Square, Suite 6-J
Lexington, KY 40507

Illinois Coal Operators Association
   117 S. Fifth Street (710 Reisch Building)
Springfield, IL 62701

Indiana Coal Association
632 Cherry Street
Terre Haute, IN 47808

Kanawha Coal Operators Association
809 Kanawha Valley Building
Charleston, WV 25301

Kentucky Coal Association
PO Box 4242
Lexington, KY 40504

Kentucky Reclamation Association
PO Box 217
Earlington, KY 42401

Kentucky River Mining Institute
PO Box 838
Hazard, KY 41701

Kentucky Independent Coal Producers, Inc.
% Coal Operators and Associates, Inc.
315 Hopkins Building
Pikeville, KY 41501

Kentucky-Tennessee Coal Operators
    Association, Inc.
1001 Circle Drive
Corbin, KY 40701

Keystone Bituminous Coal Association
311 Towne House
Harrisburg, PA 17102

League of Women Voters Mountain Plains
    Coalition
2160 Vassar Drive
Boulder, CO 80303

Logan Coal Operators Association
PO Box 240
Logan, WV 25601

Midwest Coal Producers Institute
117 S. Fifth Street, Suite 710
Springfield, IL 62701

Montana Coal Council
PO Box 1324
Dillon, MT 59725

National Ash Association, Inc. (NAA)
1819 H Street, NW
Washington, DC 20006

National Coal Association (NCA)
Coal Building
1130 17th Street, NW
Washington, DC 20036

National Council of Coal Lessors
317 Southern Building
Washington, DC 20005

National Independent Coal Operators'
    Association (NICOA)
PO Box 354
Richlands, VA 24631

New Mexico Mining Association
PO Box 5527
Santa Fe, NM 87502

Northern West Virginia Coal Association
PO Box 1386
Fairmont, WV 26555

Ohio Valley Coal Operators Association
209 Lew-Port Building
St. Clairsville, OH 43950

Pennsylvania Coal Mining Association
240 N. Third Street
Harrisburg, PA 17101

Pennsylvania State University
Coal Research Section
513 Deike Building
University Park, PA 16802

Powder River Basin Resource Council
PO Box 6221
Sheridan, WY 82801

Rocky Mountain Coal Mining Institute
356 Lafayette Street
Denver, CO 80218

Slurry Transport Association (STA)
490 L'Enfant Plaza East, SW, Suite 3210
Washington, DC 20024

Society of Mining Engineers of Aime
540 Arapeen Drive
Salt Lake City, UT 84108

Southern Appalachian Coal Operators
    Association
825 Whitehall Road
Knoxville, TN 37919

Southern Coal Producers' Association
809 Kanawha Valley Building
Charleston, WV 25301

Southern Virginias Coal Association
Bluefield, WV 24701

Southwestern Interstate Coal Operators
    Association
Bottenfield Building
Walnut and Fifth Streets
Pittsburg, KS 66762

Surface Mining Research Library
1218 Quarrier Street
Charleston, WV 25301

United Mine Workers of America (UMWA)
900 Fifteenth Street, NW
Washington, DC 20005

Utah Mining Association
825 Kearns Building
Salt Lake City, UT 84101

Virginia Coal Association
700 Building, Suite 1506
7th and Main Streets
Richmond, VA 22319

Virginia Mining Institute
Southern Division
Drawer V
Big Stone Gap, VA 24219

Virginia Mining Institute
Northern Division
PO Box 83
Tazewell, VA 24651

Virginia Surface Mining and Reclamation
    Association
PO Box 706
Norton, VA 24273

West Virginia Coal Association
1340 One Valley Square
Charleston, WV 25301

West Virginia Surface Mining & Reclamation
    Association
1624 Kanawha Boulevard, East
Charleston, WV 25311

West Virginia University
Coal Research Bureau
219 White Hall
Morgantown, WV 26506

Western Pennsylvania Coal Operators
    Association
715 Henry W. Oliver Building
Pittsburgh, PA 15222